Mobile Phone Cultures

Mobile cell phones are used all around the world, and now play an important role in the everyday lives of billions of people. What do we really know about mobile phone culture, however? This fascinating, provocative and comprehensive collection explores the cultural and media dimensions of mobile phones around the world — with an international cast of contributors and case studies from many countries across the Americas, Europe and Asia.

Mobile Phone Cultures looks at how mobiles have been imagined through advertising and social representations, traces the scripting and shaping of the technology through gender, sexuality, religion, communication style, and delves into the locations of mobile culture in modernity, urban settings, and even trans-national families. The book also offers a guide to convergent mobile phone culture, with fresh, innovative accounts of text messaging, Blackberry, camera phones, moblogging, and mobile adventures in television. Indispensable for communication, cultural, media, mobile and Internet studies scholars and students alike, Mobile Phone Cultures opens up important new perspectives on how we understand this intimate yet public cultural technology.

This book was previously published as a special issue of *Continuum: Journal of Media and Cultural Studies.*

Gerard Goggin is Professor of Digital Communication and deputy director of the Centre for Social Research in Journalism and Communication, University of New South Wales, Sydney, Australia. He has published widely on new media and culture, and his books include Internationalizing Internet Studies (2007), Cell Phone Culture (2006), Virtual Nation: The Internet in Australia (2004), and Digital Disability (2003). Gerard is editor of Media International Australia.

Mobile Phone Cultures

Edited by
Gerard Goggin

Routledge
Taylor & Francis Group
LONDON AND NEW YORK

First published 2008 by Routledge
2 Park Square, Milton Park, Abingdon, Oxon, OX14 4RN

Simultaneously published in the USA and Canada by Routledge
711 Third Avenue, New York, NY 10017

Routledge is an imprint of the Taylor & Francis Group, an informa business

Typeset in Minion 10.5/13pt by the Alden Group, OxfordShire

British Library Cataloguing in Publication Data
A catalogue record for this book is available from the British Library

ISBN 13: 978-0-415-42530-8

CONTENTS

Introduction—Mobile Phone Cultures

Gerard Goggin

With mobile phones firmly ensconced in the car, hand, and pockets, and adorning the bodies and the clothes of literally billions of people around the world, it is high time for a consideration of the cultures associated with this exemplary everyday technology.

Certainly, there is no lack of popular, policy, industrial, and now scholarly interest in mobile technologies. In 2005–2007 there has literally been an outpouring of books from the academic presses—most of which are discussed or at least mentioned in the pages that follow. The field of mobile studies is now a fascinating interdisciplinary and increasingly international one, which overlaps, and may be contrasted with, a number of other inquiries into technology, not least Internet studies. Yet while scholarship relating to mobiles has centred on questions of communication and the social in particular (if I may be permitted such a generalization), there has not yet been the sustained concentration on questions of culture that the technology, its meanings, practices, and consumption merit.

To bring together leading cultural studies of mobiles, then, is the rationale for this collection. The journal *Continuum* is a perfect venue for such a presentation because it addresses audiences and specialists in cultural and media studies. My sense is that cultural and media studies have been slower than other disciplines to approach mobiles as an object of study. Elsewhere I have suggested that this has to do with the history of telecommunications and its study (Goggin, 2006, 2007). The telephone and telecommunications have not received the same attention in cultural and media studies, as, for instance, have communications and media technologies and forms subtended by press, radio, and television, or even certain new media forms such as video and computer games, and the Internet.

At this conjuncture, however, attention to the cultural and media aspects of mobiles is something sorely needed. There is the task of analysis and theorization of mobiles cultures, recognizing their specific contexts, and then the tracing of the articulations of these with other forms of communicative and media cultures. Then there is the comprehension of mobiles cultures as part of the larger local and international logics of cultural transformation.

Just as cultural and media scholars have something quite important to add to the endeavour of understanding mobiles, they also have something to gain—over and above the imperative of rethinking their concepts, methods, and dispositions in the encounter with and construction of a new object of study. The belated arrival of cultural and media studies on the scene means that it needs to enter a dialogue with the many approaches, disciplines, and relatively well-developed understandings and assumptions that already traverse the rich field of mobile studies. This necessarily brief establishment of this context hopefully explains key features of this collection, and draws attention to some possibilities it holds.

The collection opens with three papers on aspects of 'Imagining Mobiles'. Leading Spanish scholars Juan Miguel Aguado and Inmaculada J. Martínez take an international look at the important role of advertising discourses in their 'The Construction of the Mobile Experience: the Role of Advertising Campaigns in the Appropriation of Mobile Phone Technologies'. One of the foremost scholars of mobiles today, Leopoldina Fortunati, is joined by her colleagues Alberta Contarello and Mauro Sarrica in their 'Social Thinking and the Mobile Phone', a fascinating study that takes a social representations approach to understanding how the mobile phone is conceived by its users. Catherine Middleton contributes a study of the much loved and equally reviled Canadian BlackBerry technology in her 'Illusions of Balance and Control in an Always-on Environment'.

The theme of the next section is 'Shaping Mobiles', and the four papers under this rubric offer different perspectives on this. Leslie Regan Shade investigates the striking invocations of gender that mediate mobiles in her 'Feminizing the Mobile: Gender Scripting of Mobiles in North America'. In '"What Hath God Wrought?" Considering How Religious Communities Culture (*or Kosher*) the Cell Phone', Heidi Campbell looks at the relationship between mobiles and religion. Campbell's study not only contributes to the growing literature on mobiles and religion but also—following the important work of Diane Zimmerman Umble on the telephone in Old Order Mennonite and Amish life—highlights the complexity of people's reactions to and affects arising from technology, and how constitutive this is. In her 'Pocket Technospaces: the Bodily Incorporation of Mobile Media', Ingrid Richardson offers an ethno-phenomenology of mobile media, exploring the implication of the corporeal, perceptual, and spatial in how the technology and its users are co-produced. While the work of Bruno Latour is often mentioned as orienting studies of mobiles, Chris Chesher's 'Becoming the Milky Way: Mobile Phones and Actor Networks at a U2 Concert' offers a rare, elaborated reading.

In the following section, 'Locations of Mobile Culture', the four contributions provide conceptualizations of the mobile informed by specific settings. Along with Japan, Korea has been celebrated for being in the van of mobile media use, and Larissa

Hjorth offers a sophisticated reconsideration of this in her 'Snapshots of *Almost* Contact: the Rise of Camera Phone Practices and a Case Study in Seoul, Korea'. With Tanya Batson-Savage's '"Hol' Awn Mek a Answer mi Cellular": Sex, Sexuality and the Cellular Phone in Urban Jamaica', we move to the West Indies, and how mobiles have taken centre stage in popular culture—as evidenced especially by comedy, song lyrics, music video and music hall, and street culture. The Philippines has often been celebrated for its embrace of mobiles culture, especially text messaging, and Cecilia Uy-Tioco extends our understanding of how this underpins the nation's diaspora in her 'Overseas Filipino Workers and Text Messaging: Reinventing Transnational Mothering'. Mark McLelland surveys the literature on 'Socio-cultural Aspects of Mobile Communication Technologies in Asia and the Pacific', and articulates a compelling and detailed case for research directions.

The last section of the collection, 'Mobile Image and Text', looks at the emergence of two important facets of mobiles culture that presage future developments in mobile media. In her 'Picture This: the Impact of Mobile Camera Phones on Personal Photographic Practices', Lisa Gye reviews the relation of mobiles to personal photography, and discusses the camera phone's social functions and cultural implications. Virginia Nightingale's 'The Cameraphone and Online Image Sharing' also considers mobile photography, but with a focus on cultural, affective, and gift economies of networked image exchange, and their industrial configurations. Angel Lin and Avin Tong offer a nuanced and detailed reflection on text messaging and how it is understood, informed by a suggestive study in their 'Text-messaging Cultures of College Girls in Hong Kong: SMS as Resources for Achieving Intimacy and Gift-exchange with Multiple Functions'. Finally, Christina Spurgeon and I study the intersections between mobiles and television cultures, in our 'Mobiles into Media: Premium Rate SMS and the Adaptation of Television to Interactive Communication Cultures'.

I am grateful for *Continuum* editors Brian Shoesmith, Panizza Allmark, and especially Mark Gibson for their receptiveness to this special issue topic, as well as their patience and support in seeing it to completion. I would also like to thank Katherine Burton for the opportunity to publish these studies as a book collection, something that stands to enlarge its circle of readers. I wish to acknowledge the support of the Australian Research Council for the research fellowship that has made this work possible. Finally I thank my family, Jacqueline, Liam, and Bianca, for their inestimable love and patience, as always.

References

Goggin, G. (2006) *Cell Phone Culture: Mobile Technology in Everyday Life*, Routledge, New York and London.

Goggin, G. (2007) 'Cultural studies of mobile communication', in *Handbook of Mobile Communication Studies*, ed. J. E. Katz, MIT Press, Cambridge, MA.

The Construction of the Mobile Experience: the Role of Advertising Campaigns in the Appropriation of Mobile Phone Technologies

Juan Miguel Aguado & Inmaculada J. Martínez

Introduction: Advertising Discourse and Domestication Processes

The current approach to the social uses of the mobile phone tends to emphasize their private nature, restricting the configuration of their shared image to those meanings performed in personal and group interaction rituals. The use of the idea of domestication is a good example of this prevalence given to the role of social private actors in the appropriation of the mobile phone. Domestication originally referred to the process of socio-cultural appropriation of technologies in the context of the family home (Silverstone *et al.*, 1992; Silverstone, 1994; Silverstone & Haddon, 1996). In this paper we intend to revise the term for the field of mobile phone technology appropriation by focusing on the role that commercial discourses play in the construction of a coherent imaginary of these technologies.

Two conceptual choices mark our critique of the current literature on domestication: the first is to open the appropriation process to other cultural spaces beyond the strict limits of the home, which seems almost obvious when approaching mobile technologies (Haddon, 2000). The second choice is to consider appropriation as a part of consumption processes, and, hence, as a simultaneously public and private process. Our aim is to verify the oft-presupposed complicity between institutional and non-institutional discourses and to clarify the way in which they cooperate to connect the imaginary of mobile phone technologies to the concrete 'biography of objects' (Kopytoff, 1986).

Consumption is a cultural phenomenon that increasingly involves specific forms of appropriation, re-structuration and transformation of meaning practices with regard to things and symbols (Appadurai, 1986). This approach to consumption as a meaning negotiation process seems especially relevant in the case of mobile phone technologies, in which appropriation by users must face a significant complexity derived from their strange condition with regard to standardized practices and meanings.

Following Ibáñez (1994), we understand advertising as a prevalent institutional discourse in the context of consumption practices. Through advertising, institutional actors pose and disseminate changing proposals of meaning frames attached to functional instructions, suggestions and evocative analogies. Its symbolic resources, however, are not only social contexts of use or mere plain representations of values, but life episodes, biographic anecdotes and experiential instructions. By means of advertising discourse, products, services or brands become not just things, behaviours or names that ultimately serve the purpose of satisfying needs or desires for consumers, but systems of meanings and practices that involve social rituals and discourses (Baudrillard, 1981; Appadurai, 1986; Featherstone, 2001). Advertising discourse transforms practices, objects and subjects into signs that are used as parts of a cultural conversation. In this way, products and services become the medium through which brands and consumers construct each other in terms of social interaction.

Advertising Experiences and Experiencing Identities

Understanding consumption as a meaning negotiation process between institutional and non-institutional discourses asks for a kind of 'common language' by means of which they represent—and thus recognize—each other. That coherent code allows translating the institutional discursive axis (the use of the product or service, the social image of the brand) to the social discursive axis (interaction patterns and values) and vice versa. The presentation of *recognizable experiences* related to objects, uses and lifestyles operates in our understanding as the 'common code' for institutional and non-institutional discourses.

On the side of institutional discourses, Pine & Gilmore (1999), Rifkin (2000) and Schmitt (1999) observe an experiential turn in marketing strategies. In a context of increasingly undifferentiated products and services, the influence of multi-level commercialization strategies developed in cultural industries and the adaptive potential introduced by ICTs have moved the focus of consumption dynamics from property to access (Rifkin, 2000) so that the value of consumption lies less in owning a product than in having access to its use. In such a context the assets of a brand lie rather in its capability to generate loyal customers than in its capacity to generate products. Consequently, companies turn their attention to the way in which consumers experience their products and services, codifying the access to goods or services in terms of experiences that link brand imagery to customers' social imagery through values, emotions and identity markers.

On the side of non-institutional discourses, biographies and identities (related to both objects and subjects) are codified in terms of significant experiences, that is, life events which involve meanings, emotions, uses and interactions which are registered and integrated into what Giddens (1991) calls the narratives of the self.

In the case of mobile phones, advertising discourses usually focus on the social interaction nature of the mobile experience and on the transparency of experiential mediation (accessing mediated experiences equates to experiencing them). Brand or campaign slogans constitute a good example of this: 'connecting people', 'experience life'; 'don't miss a moment', 'see new, hear new, feel new', 'the vivacity of life', or 'live it!'.

From a functional point of view, advertisements are addressed to produce discursive coherence between institutional and non-institutional discourses, becoming a sort of meeting point where advertisers' meaning proposals and social context-situated meaning practices confront their symbolic imaginaries and represent each other. This kind of discursive mediation is the key element of MacCannell's (1976) concept of cultural experiences.

Cultural experiences refer to the kinds of experiences that constitute consumption practices, and, subsequently, commercial and media discourses. According to previous work (Thompson, 1995; Ritzer, 1999; Aguado, 2002) we understand cultural experiences as social meaning constructs that link social imaginaries with life events in terms of access and consumption. An intuitive way to understand cultural experiences is to observe the way they work in tourism and leisure: a theme park or a tourist site organizes the structure, nature and time of consumers' experiences constituting simultaneously the main resource for their realization (Richards, 2001). The processes through which cultural experiences are formed involve three relevant elements: experiential frames, mediations and experiential realizations.

Experiential frames are meaning structures that evoke—and allow reproduction of—experiences in terms of meaning interactions. They operate as sense-making patterns with regard to interaction situations and include *markers* (signs and objects that communicate the performing nature of the situation), *scenarios* (arrangements of objects and spaces in order to evoke meaning structures), *demonstrations* (actions, roles and situations linked to experiences) and *meta-experiences* (instructions about the appropriate interactions with environment in order to successfully reproduce experiences).

Experiential realizations refer to the concrete experience of consumers and involve their participation in the embodiment of an experiential frame within a given situation. Experiential realizations include the *appropriation of meanings* through emotions and shared values, the *coordination of meaning patterns* in communities of practice, the *evaluation* of the experience with regard to its biographical coherence, and the *register of the experience* in order to ensure its biographical insertion, its authentication and its sharing with others.

Both concepts are inspired by the theoretical work of Goffman (1963) and Garfinkel (1967). Experiential realizations take place in situations of interaction as a part of the confluence between actors' biographies and the biography of things (Kopytoff, 1986). Experiential frames play a relevant role in the definition of interaction situations, thus contributing to build meaning nets that coordinate actions and interpretations. Experiential frames are produced and reproduced through discourses and, consequently, the encounter between experiential frames and experiential realizations always involves some form of mediation.

Mediation works as a coherence producer between experiential frames and experiential realizations. Mediation refers to any agency that (re)presents experiential frames and attributes values to them, making them recognizable to users in terms of interaction situations and identity differentiation. It is almost self-evident in this context to point to advertising discourse as a prevalent mediating agent in the context of experience consumption, and, consequently, as a relevant part of the appropriation process (Haddon, 1998).

Advertising Mobile Phones: Discursive Strategies, Experiential Frames and Experiential Realizations

In the following section we present the general outcomes of a research on the impact of advertising campaigns in the appropriation of mobile phone technologies. The fieldwork and analysis was out between June and December 2005 as part of wider research on experiential mediation and ITC appropriation developed by the Research Group on Communication, Culture and Technology at the University of Murcia (Spain). The theoretical framework and the general hypothesis follow the assumptions outlined in the preceding sections, placing emphasis on advertising discourse as a mediation agent between experiential frames and experiential realizations in the context of the commercialization of experience.

Methodology

Our aim was to look at (a) how ads represent the experience of mobile phone technology; and (b) how users perceive the coherence between their everyday use of mobile phones and their perception of the experiential frames presented in advertising discourse about mobile phones. In doing so, we compared the results of a discourse analysis on a sample of TV and press ads with a series of three discussions with mobile phone users belonging to different age groups. The sample for the discourse analysis included 25 TV commercials and 25 press ads, dating from the period 2003–2005. Eight of them were part of international campaigns and the remainder were part of national campaigns from Spain, Portugal, France, the United Kingdom, the United States, Australia, Japan and Argentina. Due to their small presence, their local nature and their limited impact on audiences (CMT, 2005), the selected sample excluded ads by mobile content providers. Consequently, the sample for the discourse analysis included TV and press ads from the period 2003–2005 by manufacturers (Nokia, Motorola, Samsung, LG, Panasonic, Sony-Ericsson) and operators (Vodafone, Orange, Movistar, Amena, Nextel, T-Mobile and Universal Mobile).

Discussions were distributed in three age groups, each of eight informants. The informants of these groups were all experienced mobile phone users. The first group included people aged 15–19. The second group comprised young adults aged 20–35. Informants who were aged 45 years or older constituted the third group. The guidelines for the group interviews were twofold: they were asked for comments on their everyday experience with mobile phones, as well as comments on a series of five TV ads and five press ads selected from the discourse analysis sample. Accordingly, the intersection between the institutional and non-institutional discourses was investigated from two perspectives: firstly, at the level of the discussion groups (on the basis of the informants' freely offered comments on ads, as in campaign post-tests); secondly, at the level of discourse analysis of both the sample of ads and the transcripts of discussions.

With regard to the sample of ads, a discriminatory pre-selection was made on the basis of two differentiating criteria: perceived advertiser's intention (the purpose of the ad) and perceived focus in the argumentative structure of ads (where the advertisers placed stress on persuasion). Three communication patterns were consequently discriminated: *competitive ads*, *narrative ads* and *illustrative ads*.

The first category includes advertisements aimed at exclusively differentiating the product or service from competitors. Competitive ads qualify objects through functional values typical of the social representation of technologies, such as ease of use, quality, comfort, safety, price, and so on. They often involve classical advertising discursive forms such as comparison, exhibition and information. Discursive representation restricts technological values to the concrete purchasing interest. Due to this contingent nature and their limited reference to the social imagery of mobile phone use, competitive ads were excluded from the sample.

Narrative ads, secondly, involve discourses organized through socially and biographically situated interaction frames. They aim to socialize the uses of mobile phone technology concerning interaction rituals, lifestyles and person-linked aesthetics in reference to both individual and collective identity. Discursive forms prevalent in these cases are performance and presentation, depending on whether the emphasis is placed on the social situation or on its conditions for viability (aesthetics or lifestyle as a precondition for successful social relations). Discursive representation resorts here to social interactions, and the mobile phone is presented, so to speak, as a 'medium with a human face'. This pattern is illustrated by Nokia campaigns that place emphasis on visually representing the brand slogan ('Connecting people') or camera phone ads that emphasize social situations in which available camera use affects the meaning of interactions.

Illustrative ads involve discourses simultaneously structured upon objects and concrete interactions, both aiming at socially explaining the uses of mobile phone technology. In this case, advertising discourse connects technology-centred functional values (design, usability, functional versatility) to consumer identity-centred values (aesthetics, identity markers, lifestyles). The stress is placed on the interaction *with* technology rather than in the social interaction *through* technology. As such, it demands a previous clear delimitation of targets in terms of consumer profile classification and standard situations in which interaction with technology makes a difference.

The kinds of ads selected for the discourse analysis sample and for the discussion groups fit both narrative and illustrative categories.

Discursive Strategies

Discursive strategies refer to the way in which texts connect products or services to social interaction and identity frames. The function of the advertising discourse is to assimilate the symbolic connotations of both the social interactions and the objects, dealing with them as interpretable signs. Assimilation works as evocative analogy by presentation of a meaning frame (connotations of the mobile phone technologies) through the symbolic resources of the other (connotations of social interactions). The experiential frames presented this way act as guiding lines for both the discursive coherence and the concrete experience of the product or service. As a result of the sample analysis and discussion group perceptions, we identified four different discursive strategies:

- Aesthetic assimilation: the experiential frame is posed through the aesthetic characterization of the mobile phone device, the user-target presented or the formal dimension of the advertising text. The mobile phone is presented as an exclusive portable product strongly attached to one's identity in terms of everyday use or as a part of one's body. Implicit experiential analogies to watches, pens, perfumes, clothing or jewels are common in these cases. Such analogies are built through evident context-of-use substitution: a picture of a phone hanging on the walls of an art museum (Samsung SGH-T500); a phone hanging on a woman's neck (Motorola V3); a phone lying with make-up, mirror and perfumes on a dressing table (Samsung A800). Also through the meaningful use of inter-textual visual aesthetics: using the visual code of perfume ads, advertisers underline the body-attached nature of the mobile phone and its value in identity presentation. Or finally through the aesthetic coherence between the user-target characterization and the object (as in Nokia's 7280 collection series).
- Biographic assimilation: the experiential frame is posed through the performance of meaningful life episodes that simultaneously involve lifestyles and recognizable context-situated social interactions. This tends to fuse social use rituals and consumption rituals, merging the emotional and interactional consequences of both. Examples of this discursive strategy include ads presenting romantic messaging situations (Nokia 7650), friends joking with ringtones (Movistar, Orange), and

people using mobile phone lights in concerts (Movistar). The stress in these cases is placed on the positive characterization of the social interaction—and less so on the functional operation, which, in turn, constitutes the base for the following discursive strategy.

- Functional assimilation: the experiential frame is posed in terms of a demonstrative exemplification. Social interaction remains in the background. The purpose of the performance is thus to instruct about the procedures of use, putting it in the context of both social and cultural values: the kinds of interactions in which it is operative, and the social and identity values derived from its use. A meaningful example of such a discursive strategy can be observed in those imaging phone campaigns in which instructive performances of socially situated use are presented. Sony-Ericsson E530 and Nokia N90 ads, for instance, emphasize the availability and ease of use of the high-quality camera function through a hyper-realistic spectacular exhibition of people capturing visually shocking events (a bear catching a salmon in a wild river, or a baby doing a somersault while playing on a trampoline). In the case of camera phone ads it is important to notice that values such as ease of use and image quality are merged in the characterization of the mobile phone as a transparent medium (Baudrillard, 1983). In this way, suggesting a desired assimilation between reality and its representation relates to the mediating social use of imaging mobile phone technology, as a kind of hyper-realization (Lasen, 2002; Rivière, 2005; Goggin, 2005).
- Metaphoric assimilation: the experiential frame is posed through visual metaphors that refer to a given key-value derived from the individual, or, to a lesser extent, social use of the mobile phone, the operator's service, or the multimedia content. The social or individual value performed by evocative analogy is substituted by the impact of spectacular representations that prefigure the added value upon which the message is built. A good example of this discursive strategy is the T-Mobile campaign about changeable monthly instalments according to the user's convenience. The campaign slogan ('you always know what is coming') is emphasized by a visual metaphoric performance in which a man walks along the street safely anticipating unexpected events (a piano falling from a moving work, a car crashing through a shop window, etc.). Advertising discourses on mobile phone content (games, music, ringtones, videos, and news) also tend to resort to visual metaphors in order to represent the experience derived from their consumption. Beyond some lifestyle connotations, social interaction context remains obviated. Conversely, this discursive strategy tends to underline individual, emotional or even sensorial experiences that present a conception of the mobile phone close to that of nomadic privacy devices such as Walkmans, MP3 players, and game consoles (Araujo de Souza, 2004).

Mobile Phone Experiential Frames and Realizations

According to the sample analysis, advertising discourses distribute mobile phone experiential frames in four significant levels shown to partially overlap with the four differentiated discursive strategies presented above. Most of the connotations derived from these four levels are easily recognized as relevant marks of the mobile phone social imagery, consequently finding their correlate in focus group discourses:

- Identity experience: the mobile phone device is a part of one's body and, consequently, it is also an important part of the aesthetics of one's personal identity, both in its interactional and visual dimensions. Seemingly corresponding to previous research (Ling, 2001, 2004; Scifo, 2005), discussion participants from groups one and two as well as qualified professionals in groups three and four were shown to be more sensitive to experiencing their mobile use in terms of identity markers. However, the vast majority of participants agreed that they could not live without their phones and that, after its functionalities, the look of the mobile phone was relevant for them. Finally, a general consensus on the increasing similarities between mobile phones and identity-attached portable goods, such as watches, was reached. Both in advertising discourses and in discussions it was not only the physical aspect of the device that was related to identity presentation rituals; the way it is used in social interactions was also considered relevant in regard to lifestyles and role assumption.
- Control experience: the mobile phone is an inclusive and always-available technology device. In this sense, the mobile phone is experienced as a sort of 'meta-device', a device that includes a set of integrated technology devices with multi-level functionalities. Though conversation is still a highly relevant function, the mobile phone is no longer a 'phone'. With the exception of two participants in group four (aged 56 and 62), all of those interviewed admitted to intensive use of functions other than conversation. The younger members of groups one and two absolutely refused to consider a hypothetical portable voice-connection device as a mobile phone. Users experience the inclusive and available multi-functional condition under the same logic of control that characterizes other integrated portable multi-functional devices. At a metaphoric level we could say that mobile phones are like Swiss Army knives: they constitute a portable repository of functional resources that allow control of unexpected situations or events. The semantic connection between design and multi-functionality, which characterizes advertising discourses in the sample directly, relates here to the users' experiential connection between the mobile phone's identity relevance and functional capabilities.
- Social group experience: the mobile phone is essentially a social tool through which one can experience group interactions in new or more intense ways. Newness and intensity are the common values that merge the emotional implications of mobile phone social connotations. Mobile phones provide users with new ways of interacting—images, instant messaging, group chats, sound clips, or group calls—as well as enhancing standard social interactions through availability, instant coordination, or virtual permanent presence: 'through a single ring I'm telling I'm with him' (Cristina, 24). To the role of affective and emotional dimensions of social group experience, advertising discourse adds a dimension of professional interaction, where the same values of newness and intensity are related to efficacy and comfort.

The vast diversity of social interaction situations performed in mobile telephone ads is linked mostly to life episodes: at the beach with friends, the giving of Christmas presents, at the office with colleagues, at the airport for business travel, at home with children. They fulfil the double role of user instructions and socialization. The way in which these interactions are presented emphasizes the (positive) social consequences of mobile phone use, building a specific way of experiencing them. Thus, when asked whether they could imagine a personal life event that could constitute the basis of a hypothetical ad on mobile telephony, the majority of participants in the discussions answered affirmatively, posing a variety of personal anecdotes that, in essence, fit the nature of advertising discourses.

- Self-experience: the mobile phone is a medium that relates to one's everyday experiences. As such it mediates one's own access to experiences as well as those of others. As a nomadic privacy device, on the one hand, the mobile phone allows access to preformed sensory and meaning experiences in the very same terms that mass media do, adding the value of ubiquity. We could speak at this point of an incipient process of mobile phone mediatization, in the sense of an increasing convergence between mobile phone social uses and mass media content consumption standards. On the other hand, as a suitable tool for registering and managing one's everyday experiences (Ling, 2004; Rivière, 2005; Goggin, 2005), the mobile phone is posed as a medium for producing or enhancing aspects of self-perception. This includes vicarious access to mediated experiences and sharing them with others as a meaning resource in social interactions.

The use of camera phone images deserves special mention at this point. In the context of mobile phone technologies, digital images do not function as a mere memory record. Their attached-to-the-moment condition and their manipulability give them an added dimension: as an ontological proof that validates the occurrence of experienced events, and, simultaneously, as a semantic resource that can be re-contextualized in different symbolic environments. Advertising discourses on imaging phones resort to these aspects by emphasizing the ability to capture unexpected or unusual moments and the diversity of uses of the generated images (Koskinen & Kurvinen, 2005). Accordingly, the self-experience frame is deeply correlated to the identity experience frame in advertising discourses and in the perception of users. Again, it is the younger participants in groups one and two who, depending on the relevant role that the mobile phone plays in the presentation of identity, to a larger extent than their older counterparts, internalize this experiential frame.

Conclusions: The Mobile Phone as a Cultural Experience

The empirical evidence extracted from the discourse analysis of television and press ads and group interviews suggests a significant coherence between institutional and non-institutional discourses as a relevant part of mobile phone appropriation processes. An additional interesting outcome is that, while findings derived from the discussion groups broadly fit previous work on the social uses of mobile telephony (Katz & Aarhus, 2002; Ling, 2004), so too does the advertising discourse analysis. This coherence allows the presupposing of a kind of symbolic exchange between institutional and non-institutional discourses that denies the prevalent private nature of mobile phone technology appropriation in current research, demanding a revision of the idea of domestication as a conceptual resource. The meaning context of this interdependency between private and public discourses is shown, in our view, in the form of a mediated construction of experiential frames.

Consequently, advertising discourse is not only about selling products or services, just as consumption is not only about buying or using products and services. People gather media discourses and transform consumption proposals into meaningful social practices. In their turn, campaign designers gather social practices and meanings from everyday life in order to adapt them to the experiential imagery of the product or brand, thus contributing to the production of new meanings and practices. The role of advertising and its convergence with other discourse forms in the cultural industries (such as entertainment or news) constitutes, then, a key element in the process of constructing a cultural imaginary of the mobile phone.

Furthermore, both discourse producers—advertisers and users—are strategically aware of this coherence. The permanent reference in advertising to social interaction frames is understood by users as a strategic selection and representation of their own discourse on the social use of mobile phone technologies. In other words, interviewed users think that what ads say about mobile phones is essentially coherent with their experience of them, taking for granted that advertisers represent social experiences for the purpose of selling.

In this sense, advertising discourses about mobile phones do not deal mainly with technologies, prices or services, but with emotions (which are transmitted and shared), social situations of interaction (related to security, availability, and personal identity), individual and collective experiences (fun, romance, and friendship) or technology-related values (such as efficiency or design). The socio-cultural impact of mobile phone technologies is always implicit in the advertising discourse as a matter of mediation between the concrete experience of users and the experiential frame implemented by technology designers. This mediation operates in both senses, providing instructions for the appropriate use of meanings and objects, and inserting in the cultural imagery of the commercialized object or service those social symbolic and instrumental transformations operative in communities of meaning and practice.

References

Aguado, J. M. (2002) 'La mediación tecnológica de la experiencia: la globalización de los marcos experienciales en la construcción de imaginarios socioculturales', *Razón y Palabra*, vol. 27. Available at: http://www.cem.itesm.mx/dacs/publicaciones/logos/anteriores/n27/jaguado.html.

Appadurai, A. (1986) 'Introduction', in *The Social Life of Things*, ed. A. Appadurai, Cambridge University Press, Cambridge, pp. 3–63.

Araujo de Souza, A. (2004) *From Multiuser Environments as (Virtual) Spaces to (Hybrid) Spaces as Multiuser Environments: Nomadic Technology Devices and Hybrid Communication Places*, PhD Thesis, Universidad Federal de Río de Janeiro, Brazil.

Baudrillard, J. (1981) *For a Critique of the Political Economy of the Sign*, Telos Press, St. Louis.

Baudrillard, J. (1983) *Simulations*, Semiotexte, New York.

CMT (Comisión del Mercado de las Telecomunicaciones) (2005) *Informe del sector de la telefonía móvil en España, 2004*, CMT, Madrid.

Featherstone, M. (2001) *Consumer Culture and Postmodernism*, Sage, London.

Garfinkel, H. (1967) *Studies in Ethnomethodology*, Prentice Hall, Englewood Cliffs, NJ.

Giddens, A. (1991) *Modernity and Self-identity; Self and Society in the Late Modern Age*, Polity Press, Cambridge.

Goffman, E. (1963) *Interaction Ritual: Essays on Face-to-face Behavior*, Anchor Books, New York.

Goggin, G. (2005) 'Calling the shots', *The Age*, 2 Jul., Available at: http://www.theage.com.au/

Haddon, L. (1998) 'The experience of the mobile phone', Paper presented at the *XIV World Congress of Sociology*, 'Social Knowledge: Heritage, Challenges, Prospects', Montreal, 26 Jul.–1 Aug.

Haddon, L. (2000) 'Domestication and mobile phone telephony', Paper presented at the 'Machines that Become Us' conference, Rutgers University, New Jersey.

Ibáñez, J. (1994) 'Una publicidad que se anuncia a sí misma', in *Por una sociología de la vida cotidiana*, ed. J. Ibáñez, Siglo XXI, Madrid, pp. 129–153.

Katz, J. & Aarhus, M. (eds) (2002) *Perpetual Contact: Mobile Communication, Private Talk, Public Performance*, Cambridge University Press, Cambridge.

Kopytoff, I. (1986) 'The cultural biography of things: commoditization as process', in *The Social Life of Things*, ed. A. Appadurai, Cambridge University Press, Cambridge, pp. 64–94.

Koskinen, I. & Kurvinen, E. (2005) 'Mobile multimedia and users: on the domestication of mobile multimedia', *Telektronikk*, no. 3, pp. 60–68.

Lasen, A. (2002) *A Comparative Study of Mobile Phone Use in London, Madrid and Paris*, Digital World Research Centre, University of Surrey, Available at: http://www.surrey.ac.uk/dwrc/Publications/CompStudy.pdf

Ling, R. (2001) *Adolescent Girls and Young Adult Men: Two Sub-cultures of the Mobile Telephone*, R&D Report R-34/2001, Kjeller, Telenor Research and Development.

Ling, R. (2004) *The Mobile Connection: the Cell Phone's Impact on Society*, Elsevier, San Francisco.

MacCannell, D. (1976) *The Tourist: a New Social Theory of the Leisure Class*, University of California Press, Berkeley.

Pine, J. P. & Gilmore, J. H. (1999) *The Experience Economy*, Harvard Business School Press, Boston.

Richards, G. (2001) 'The experience industry and the creation of attractions', in *Cultural Attractions and European Tourism*, ed. G. Richards, CAB International, Wallingford, pp. 55–69.

Rifkin, J. (2000) *The Age of Access*, Penguin Putnam, Washington, DC.

Ritzer, G. (1999) *Enchanting a Disenchanted World: Revolutionizing the Means of Consumption*, Sage, London.

Rivière, C. (2005) 'Seeing and writing on a mobile phone: new forms of sociability in interpersonal communications', in *Proceedings of Communications in the 21st Century: the Mobile Information Society*, Budapest.

Schmitt, B. H. (1999) *Experiential Marketing*, Free Press, New York.

Scifo, B. (2005) 'The domestication of the camera phone and MMS communications: the experience of young Italians', in *A Sense of Place*, ed. K. Nyíri, Passagen, Vienna, pp. 363–374.

Silverstone, R. (1994) *Television and Everyday Life*, Routledge, London.

Silverstone, R. & Haddon, L. (1996) 'Design and the domestication of information and communication technologies: technical change and everyday life', in *Communication by*

Design: the Politics of Information and Communication Technologies, eds R. Silverstone & R. Mansell, Oxford University Press, Oxford, pp. 44–74.

Silverstone, R., Hirsch, E. & Morley, D. (1992) 'Information and communication technologies and the moral economy of the household', in *Consuming Technologies*, eds R. Silverstone & E. Hirsch, Routledge, London, pp. 13–31.

Thompson, J. B. (1995) *The Media and Modernity: a Social Theory of the Media*, Stanford University Press, Palo Alto, CA.

Social Thinking and the Mobile Phone: a Study of Social Change with the Diffusion of Mobile Phones, Using a Social Representations Framework

Alberta Contarello, Leopoldina Fortunati & Mauro Sarrica

Entering the Field

One evening, you are sitting at home watching a repeat of a series produced in the late 1980s, with that particular atmosphere and the fashions of the day. One of the characters, let's say an executive, receives a phone call in his car and you suddenly realize that he is holding a type of phone that you had almost forgotten about: a sort of walkie-talkie, dark, heavy, and probably unreliable. Yet not so long ago that kind of mobile phone was at the cutting edge of high technology. Entering the field of social research into mobile phones, one has more or less the same impression. In about a decade research in this field has flourished, opening new directions of inquiry and addressing completely new issues. It is odd to read the analyses, predictions and concerns regarding what was then a strange new object.

Our brief overview gives some idea of the various directions taken by social research into the mobile over the past decade. A first focus of interest, which arose with the introduction of mobile phones on a massive scale, regards the domestication of the mobile in everyday life. Studies that fall within this macro-area aimed at reconstructing the uses and practices of the mobile in the domestic sphere, in the workplace and in the broader public arena. In this connection, we cannot but begin with the book edited in 1998 by Fortunati and the research report edited by Haddon (1998). The former examined attitudes, opinions and practices of the mobile phone (along with the telephone, answering machine, fax, computer, radio and TV) in five European countries (France, Germany, Italy, Spain and the United Kingdom). In the latter, members of the COST action 248, 'The Future European Telecommunications User', focused explicitly on histories, markets, modes of use and misuse of the mobile.

In the following year, Richard Ling edited a special issue of *Personal and Ubiquitous Computing* on 'Mobile Communication and the Reformulation of the Social Order' in which he included the new users of the mobile. Since it was no longer limited to work purposes, it became important to explore the availability of mobile communication (Licoppe & Heurtin, 2001), its integration into everyday life (Palen *et al.*, 2001), its gender use (Ling, 2001b), its negotiating use (Weilenmann, 2001) and configurations of the mobile users (Green *et al.*, 2001).

The widespread use of this device in society poses the question of the link between new users and new practices. This is perhaps why so much attention has been focused on children and youth in more recent years. Among other contributions, in 2002 a special issue of *Revista de Estudios de Juventud* on 'Juventud y teléfonos móviles', edited by Santiago Lorente, comprised a series of studies devoted to the use of the mobile by children, adolescents and youths in 10 countries. *Mobile Kommunikation: Perspectiven und Forschungsfelder*, edited by Höflich & Gebhaedt (2005), was the first volume on the mobile to address a German audience. It included the first ethnographic research on urban routes of mobile communication, pedagogical and psychological features of communication, practices and uses of the mobile within families.

At the same time, this new device demanded an interpretation at a different level. More than a simple technical device, the mobile was already stimulating the social imagination in relation to its nature and characteristics as a machine particularly close to the human body. Along these lines, contributions by Leopoldina Fortunati and James Katz focused first on its ambiguous image (Fortunati, 1998a) and identity (Fortunati, 2001), and later on relations between the mobile and the human body, also at the epistemological level, including the role of fashion and the consequences for health and well-being (Fortunati *et al.*, 2003; Fortunati, 2005). This orientation continued and focused on the machinization of the human body by means of ICTs and especially the mobile, and examined the meaning of the increasingly intimate relationship between users and ICTs (Katz, 2003; Fortunati & Contarello, 2005).

A third main level of analysis is the societal one. Under this heading we find studies that investigated the broader social changes linked to the introduction and adoption of the mobile. In this category the work of Katz & Aakhus (2002) discusses the role of the mobile at the social level, with results from many more countries. The volume edited by Brown *et al.*, (2002) is also important for the sophisticated level of analysis on the adoption and diffusion of this device, as is the work of Howard Rheingold (2002) on the creative and political use of the mobile phone to enhance mass-mobilization and democratization processes in more than one country.

Focusing on society, the debate has touched upon much broader themes. Communicative dissonances and assonances, the role of the mobile in communities and networks, the relation of the mobile to other traditional (e.g. TV) and less traditional media (e.g. the Internet) are only a few of the themes touched on in Nyíri (2003). Particularly relevant to our concerns is the proposal to interpret the situational fluidity following the spread of the mobile in terms of new 'floating worlds' (Gergen, 2002, 2003) and new—but also spiralling backwards—nomadic approaches to relationships (Meyrowitz, 2003). Moreover, in Glotz *et al.*, (2005), the problem of the meaning of the mobile is combined with cultural identities and mobile personalities, and industrial perspectives are addressed. Recently, Ling & Pedersen (2005) have analysed how mobile phone users have completely changed both the backstage and front stage of social communication that is what is spoken on the public sphere and how it is prepared and elaborated behind the scene. This book enlarges the debate on the mobile to themes which are often involved, such as language, and the psychological dimensions of mobile communication.

A further body of research includes those contributions aimed at opening a window on the future. In this group we can include works that address the further technical developments of the mobile and broaden attention to developing countries and the globalized word. In this group we can include the volume edited by Kim (2005), which collects various interesting essays on the adoption and diffusion of the mobile in developing countries such as Russia, India and the Philippines, and on intercultural aspects of practices and uses of the mobile (e.g. the construction of a diasporic identity). In Norway an entire issue, entitled 'Future Mobile Phones', of the journal *Telektronikk* was devoted to the mobile, in which it gathered interesting articles on new services and functions of the mobile, and on important turning points in the mobile's future development. We would also like to mention Haddon *et al.*, (2005), on the last generation mobiles, and Law *et al.* (2006). Finally, Nyíri (2005) revives a long-standing topic linked with the debate on ICTs—that of place—relating it to new scenarios opened up by globalization.

On the whole, this decade of studies on mobile phones looked in particular at their domestication in everyday life at a behavioural level. In other words, they studied how the use of the mobile phone has changed and reshaped many aspects of society. What was less considered was the domestication of the mobile phone at a socio-cognitive level. How has the mobile phone been integrated in shared knowledge and social thinking in everyday life? This is the research question from which we started the present empirical study.

The Present Study

To answer our research question we focused our attention on the rhythms and dynamics of the conceptual integration of the mobile phone through time. While expert debate has touched on these topics, everyday talk has not only mirrored it but also developed shared meanings about the mobile: what the risks are; what opportunities it offers; but also what images and symbolic forms are linked to it. Given this purpose, both at a methodological and a theoretical level, our choice was directed towards a social representations approach.

By means of a varied methodology which intertwines qualitative and quantitative strategies, the theory of social representations is able to grasp various levels at which social thinking structures itself and develops, as well as the processes underlying the organization of social thinking. As regards the former, we mean the content, the attitude, and the field of representation which gives shape to the representation. As regards the latter, we refer to objectification and anchoring, the social psychological processes through which people make familiar what is new, classifying and naming it, and transform what is unfamiliar into new 'reality'.

The social representations approach allows us to explore how everyday talk evolves in order to cope with highly debated novelties, charged with social and emotional aspects. It 'defines socially relevant phenomena and objects not according to any traits inherent to the objects, but according to the relation that exists between people and the objects and the events represented by them ... It is the meaning of the things in, and for, the lives of people which makes them a social object' (Wagner & Hayes, 2005, p. 122). Then it is particularly suitable for the study of the ways in which this technological device has been given 'meaning', with the passage of time, through shared communication using linguistic, iconic and metaphorical features.

Although the mobile is no longer a novelty, it has been and is the object of debate among experts and in everyday life. Thus, together with ICTs in general, mobiles have all the necessary qualifications for the social representations approach: they constitute an issue, widely shared and contextualized in a specific context (Farr & Moscovici, 1984), and in the presence of a high degree of polymorphism, i.e. with diverse features arising from different groups which give them specific values (Moliner, 1993).

The social representations perspective has been developed by several scholars who have devised methods and procedures well suited to the examination and illustration of contents and structures. Various topics have been submitted to analysis, mainly in the fields of health, education, and science. In this connection, we can refer to studies on social representations of the computer (Flick, 1996), biotechnologies (Wagner *et al.*, 2002), the Internet (Capozza *et al.*, 2003), and telecommunications (Fortunati & Manganelli, 2007). In previous research, we ourselves considered this theoretical framework well suited to the monitoring of ICTs, mainly the mobile and the Internet, with particular attention on their promises and perils with regard to body-to-body communication and to technologies of health (Contarello & Fortunati, 2006), as well as to health-related issues and perceived social well-being (Contarello & Sarrica, 2005). We also explored similarities and differences in the ways in which the mobile and the Internet are perceived in everyday life (Fortunati & Contarello, 2002) and general versus particular trends in representations and practices within five European countries (Italy, the Netherlands, Romania, Russia and Spain) (Contarello *et al.*, 2005).

The present contribution draws together data collected from the period 2001–2003 in order to explore changes in social representations of the mobile across a time span that is not wide but crucial in the emergence and dissemination of the mobile and its use. Although we have not performed a longitudinal study, we think that our cross-sectional collections enable us to perform a comparison. For this reason, we have extracted the most homogeneous samples of respondents from our previous studies and have concentrated our attention on the underlying structure of meanings associated with the mobile over time. In this way, our purpose is to contribute to the general debate with an empirical analysis of changing views of the mobile expressed not by experts but by young people in their everyday context.

Method

Participants

More than 300 psychology students at the University of Padova, Italy, were involved in the present study: 100 students (90 women, 10 men) examined in 2001, 122 (97 women and 25 men) in 2002; and 101 (69 women and 32 men) in 2003. Their mean age was 21.63. All of them had a mobile phone, except one respondent in 2001 and one respondent in 2003.

Procedure

Following Moscovici (1961/76) and Le Bouedec (1984), we considered three different components of the representation, namely information, attitude, and field of representation. Drawing inspiration from research by Di Giacomo (1980), we made extensive use of the free association technique, asking the participants to write—first, on a white sheet of the questionnaire[1]—'what comes to their mind' when presented with the key stimulus: the mobile.

Data Analysis

We explored the different components of the representation—information, attitude, representational field—through a qualitative/quantitative methodology. Data from the free association exercise were analysed through analyses provided by the Spad.T package (Lebart *et al.*, 1989). This software specializes in textual analyses, more specifically for: computing frequencies of lexical forms, finding relations between these forms (in our case through correspondence analysis), and finding relations between these forms and external variables. First, we worked on the associated words, exploring the meanings conveyed by the respondents to the stimulus. As is usual in this kind of analysis, we retained nouns, adjectives, adverbs, and short sentences. Second, synonyms and antonyms were reduced and a minimum frequency threshold was imposed. Then, we turned our attention to the distinct lexical forms of each year, through the Mocar (Mots Caractéristique) procedure. This procedure allows scrutiny of the lexical forms significantly associated by the participants in one year in comparison with the other years. Given that our objective was to monitor the changes that occurred through time, the Mocar procedure was the most suitable instrument in trying to answer our question; that is to select only those lexical forms that showed differences in citations in the three years. Third, the data were submitted to correspondence analysis (Benzécri *et al.*, 1973), an exploratory technique that is especially suited to categorical data, which allows the joint presentation of the results relating to two (or more) sets of data (Doise *et al.*, 1993; Lebart & Salem, 1988). The participants × words matrix, referring to the core stimulus, mobile, was analysed by means of the Aspar procedure in order to explore the emerging representational field and the role played by time, in terms of year of data collection (considered as illustrative variable).[2]

Results

The analysis of associations most frequently given to the mobile in the whole data collection indicates the following: SMS ($n = 106$); social relations (94); communication (89); useful (63); telephone (45); convenience (38); annoyance (38); cost (34); innovation (33); harmfulness (29); reachability (24); fashion (21); necessary (19); dependency/addiction (18); ring (17); prevents communication (15). Thus, most frequent is the association with a mode of mobile communication, SMS, which is a user design variant, launched by adolescents and youths, and which has reshaped the world of mobile communication. Next, the mobile phone is associated with the concept of usefulness in connection with social relations and, in particular, with their communicative aspects. This set of associations might refer to a support role played by the mobile phone with regard to interpersonal links and social bonds.

The overall analysis shows 993 lexical forms, 448 of which are distinct (45 per cent), and which, after treatment of the text, were reduced to 51. The 51 × 316 contingency table was submitted to the Mocar procedure offered by the Spad.T package in order to detect lexical forms characteristic of each year. We could thus further explore the content of the representation in the three different phases of our data collection.

As can be seen in Table 1, in 2001 we mostly find signs of a concerned evaluation (with reference to annoyance, futility, harmfulness) of an object which is perceived as widely shared and evokes some alarm in terms of potential addiction. In 2002, description of

Table 1 Typical associations in the three years

	Word associated	Percentage Internal	Percentage Global	Frequency Internal	Frequency Global	v-test	$p <$
2001	1 Human body[a]	2.96	1.32	9	12	2.680	0.004
	2 Annoyance	6.91	4.17	21	38	2.673	0.004
	3 Harmful to health	5.26	3.18	16	29	2.271	0.012
	4 Common	0.99	0.33	3	3	1.788	0.037
	5 A lot	0.99	0.33	3	3	1.788	0.037
	6 Dependency	3.29	1.98	10	18	1.727	0.042
	7 Futile	1.32	0.55	4	5	1.695	0.045
2002	1 SMS	16.93	11.64	54	106	3.481	0.000
	2 Telephone	7.84	4.94	25	45	2.739	0.003
	3 Display	1.88	0.99	6	9	1.618	0.053
2003	1 Necessary	4.51	2.09	13	19	3.103	0.001
	2 Useful	10.76	6.92	31	63	2.890	0.002
	3 Changes communication	1.74	0.55	5	5	2.739	0.003
	4 Invasive	2.78	1.32	8	12	2.231	0.013
	5 Reachability	4.51	2.63	13	24	2.122	0.017
	6 Sometimes	1.04	0.33	3	3	1.861	0.031
	7 Handy	2.78	1.54	8	14	1.734	0.041

[a]One aim of the 2001 research was to explore the representation of ICTs in relation to the human body and health-related topics. The human body was present in the pamphlet as one of various key stimuli.

the object comes to the fore, with particular stress on its quintessential service—SMS—while at the forefront in 2003 are references to its functional characteristics, mainly in terms of positive nuances (necessary, useful) and appreciation of it as an artefact (handy, coloured), and to the social consequences of its use, mainly in terms of invasive reachability, as well as an acknowledgement that the device had changed interpersonal communication.

The same categories × participants (51 × 316) matrix was submitted to correspondence analysis. In the illustration of the results we retained lexical forms with an absolute contribution >2.[3] On the basis of the eigenvalues histogram, we illustrate the first five dimensions (Table 2), which account overall for 16.91 per cent of the total inertia.[4] The first factor gathers at one pole—mainly due to the 2002 participants—associations with the telephone and its parts (display, logos), the specific service it offers—SMS—and social networks, i.e. close networks (friends, girlfriends/boyfriends, parents, collected in the social relations category). At the other pole, which significantly regards the 2003 research, the mobile evokes an evaluation of the functional and aesthetic features of this device (it is useful and necessary on one hand and beautiful, handy and fashionable on the other) and the social consequences of its use (dependency, invasiveness, annoyance). These consequences are seen as mainly negative, but in any case the mobile phone's ability to change the means of communicating is recognized. On the whole, this first factor shows a descriptive versus evaluating dimension which enhances reference to 'What the mobile is' and 'What it is for' versus 'How it is' and 'What it produces'. The second dimension presents at one extremity terms which refer again to the tangible aspects of this artefact (such as mast, display), especially its twofold nature: harmful but useful for emergencies; and finally new concepts: the global framework in which mobile communication is developing and a menace to the effectiveness of communication. The other pole gathers reference to the material body of this technological artefact, with its logos, rings, and again its annoyance value with regard to social relations. The opposition here seems to be between, on the one hand, the 'Public Sphere'—evoking a wide, expansive perspective—and on the other the 'Personal Sphere', more local and circumscribed. The third factor opposes opportunities offered by the device, in terms of a useful, necessary and beautiful instrument to support changes of communication and social relations, versus a good chance of freedom, but conditioned by annoyance, superfluity and, most frightening, risks of addiction which appear to be the shadow-side of a new liberty that transforms itself into forced reachability. We propose to name the factor 'The Beauty of Novelty' (which pertains especially to 2003) versus 'The Price of Freedom' (which regards the research data collected in 2001). Freedom is again at the forefront at one extremity of the fourth dimension, this time jointly with positive cues such as beauty, innovation, communication, and safety, with the downside of some annoyance. We interpret this pole as 'A Hit Bit Technology'. At the other pole, elements of uselessness, invasiveness, fashion, and commonness come together and give negative connotations to the idea of changing ways of communicating, hence the title we propose: 'A Trivial Optional'.

Lastly, the fifth factor opposes what we interpret as 'A Pleasant Gadget', where beauty is heavily impoverished by the fact that it is no longer a sign of distinction, since everybody has one (the 2002 participants situate themselves on this pole) versus 'An Effective but Expensive Technology', where the acknowledgement that the mobile phone is a handy tool that has changed modes of communication goes hand in hand with a clear recognition of its costs (expressed mainly by the 2003 respondents)

Discussion and Conclusion

Three years is not a long time, but because of the increasing rate of market penetration by the mobile, we can explore the changes which took place in time in the social representations of the mobile phone and advance some preliminary conclusions.

The three collections of data illustrated here support a particularly ambivalent representation of the mobile, which is the stable element recurring in all the social representation studies of the mobile phone starting from 1996 (Fortunati & Manganelli, 2007). Other ICTs also present elements of ambivalence, but not to the same degree as the mobile, which is really twofold. The social representation of the mobile as it emerges in our three collections of data shows a twofold division.

On the one hand, there is considerable interest in the material nature of the mobile as an artefact, with its strong aesthetic and fashion values. On the other, there is awareness of its powerful social function, which again goes in two directions: it changes and

Table 2 Correspondence analysis: first five factors

Dimension (inertia)	Words	Coordinates	Absolute contribution	Words	Coordinates	Absolute contribution
	What it is–What it is for			**How it is–What it produces**		
$I = 3.75\%$	SMS	−0.79	11.30	Uselessness	2.21	9.09
	Telephone	−1.08	8.93	Useful	0.85	7.70
	Display	−1.97	5.91	Beautiful	3.46	6.11
	Social Relations	−0.59	5.61	Necessary	1.32	5.64
	Logos	−1.66	5.59	Handy	1.19	3.34
				Fashion	0.86	2.65
				Dependency	0.91	2.54
				Invasive	1.11	2.52
				Annoyance	0.62	2.45
				Changes communication	1.67	2.36
	2002			2003		
	Public Sphere			**Personal Sphere**		
3.45%	Mast	−4.18	19.41	Logos	1.37	4.15
	Harmful to health	−1.75	16.48	Social relations	0.49	4.15
	Display	−2.70	12.14	Ring	1.01	3.20
	Rapidity	−2.43	7.62	Annoyance	0.57	2.32
	Globalization	−2.86	4.53			
	Emergency	−1.32	2.90			
	Instrument	−1.40	2.89			
	Prevents communication	−0.97	2.59			
	The Beauty of Novelty			**The Price of Freedom**		
3.35%	Beautiful	−7.71	33.88	Freedom	3.79	10.93
	Changes communication	−2.87	7.85	Annoyance	0.91	6.03
	Necessary	−1.16	4.90	Dependency	0.94	3.04
	Social relations	−0.50	4.44	Uselessness	1.17	2.88
	Useful	−0.44	2.31	Fashion	0.79	2.50
				Reachableness	0.68	2.09
	2003			2001		

(continued)

Table 2 – *Continued*

Dimension (inertia)	Words	Coordinates	Absolute contribution	Words	Coordinates	Absolute contribution
	A Hit Bit Technology			**A Trivial Accessory/Optional**		
3.22%	Freedom	−0.444.61	16.82	Uselessness	2.99	19.46
	Beautiful	−0.444.64	12.77	Invasive	2.05	10.02
	Innovation	−0.440.82	4.42	Fashion	1.17	5.70
	Annoyance	−0.440.70	3.64	Changes communication	2.20	4.80
	Communication	−0.440.44	3.37	Common	2.82	4.73
	Safety	−0.441.31	2.03			
	A Pleasant Gadget			**An Effective but Expensive Technology**		
3.14%	Beautiful	−0.447.78	36.91	Changes communication	3.42	11.89
	Uselessness	−0.441.66	6.15	Handy	1.13	3.62
	Display	−0.441.54	4.35	Always	1.24	2.80
	Fashion	−0.440.93	3.72	Social relations	0.35	2.29
	Common	−0.442.32	3.29	Cost	0.57	2.22
	Ring	−0.440.90	2.81			
	2002			2003		

enhances communication, reachability and therefore social relations, but it also prevents them, provides help in an emergency, is useful and necessary for communication but is also harmful, invasive and annoying, possibly addictive and involves high costs. All this should be placed in a global framework (whereas traditional telephony has always been local) with innovative development (Meyrowitz, 2005; Nyíri, 2005), along with a certain devaluation since the mobile phone's popularity has transformed it into a mass commodity. This picture confirms and enhances the results of research carried out on the sense of place reported by Meyrowitz (2005), on the centrality of the artefact collected by Glotz *et al.* (2005), and on the communicative dissonance and assonance studied by Gergen, who stresses new forms of absent-presence within relationships taking place in a floating world (2002, 2003).

As regards the changes registered, we note how in the course of these three years the focus moves from concerns and perceived threats related to the mobile (2001) towards an interest in the artefact in itself and on its aesthetic valence, to return in 2003 to consider how communication modes within social relations change through the mobile, with new concern as to its costs.

Over time, the entry of the mobile has evoked a twofold set of associations in everyday life. Even in the more recent data collection, as we saw, we find the presence of both positive and negative features. Interestingly enough, we find reference to new modes of communication, sometimes seen with pleasure (third factor), sometimes with concern in terms of cost (fifth factor), sometimes linked to ambivalent views (first factor). Whether these results indicate a spiralling, sinusoidal path or a simpler trajectory—from evaluation, to description, to recognition of services and functions in everyday life, with a progressive shading of symbolic charge—only further research over a longer time span can tell.

On the whole, however, we think our research study supports our initial claim. The theoretical framework and methodology we adopted seem suitable for our purpose, i.e. to explore the integration of the mobile phone at a socio-psychological level, in order to complete the framework of what some sociologists have called the 'domestication' of ICTs (Silverstone & Hirsch, 1992), and which, with Moscovici (1998), we may view as a 'metabolization' at a social-psychological level of the novel, the unusual, the unknown. For present purposes—the analysis of social change—we should, however, admit some shortcomings and give a warning. Our participants are far from constituting 'representative samples'—of Italians, young people, students.

Social representation theory and methodology proved to be suitable for our purpose, which was to study the domestication process of the mobile phone from a socio-cognitive viewpoint. However, we have to admit that the iconic and imaginative aspects foreseen by the theory are only weakly present in our data. This might be due to the particular object we analysed—the mobile phone and, by extension, technology—but it might also be a reflection of some limitation in the use of qualitative/quantitative analyses of free associations. However, the study of the integration of the mobile phone in social thinking has been able to register stability and changes that this integration has produced during the enquired period of time. In particular, it has shown changes in social thinking in relation to the mobile phone. The mobile passed from being perceived mainly as an object of material culture, defined in positive and negative terms, to both a functional and symbolic object in the following year, to a social object in the final year.

Also, this research, in common with much other research, shows the specific double face of the mobile phone, which on the one hand is considered a welcome opportunity and on the other contributes to making life harder. This double face is generally typical of all ICTs (Geser, 2004), but the radicality with which this twofold character of the mobile phone reveals itself appears peculiar to this device.

A research agenda for further studies should design studies with larger samples (Spad.T specifically answers this purpose). However, for a free association task to gather keywords related to an issue, i.e. the noise of everyday life in cafés and meeting places, as indicated by social representations theory, 100 participants would constitute a fair number. Indeed, when we moved from the numbers in the present study to almost 600 records in our international study (Contarello *et al.*, 2005), we obtained more stable and statistically reliable data, but not necessarily richer and wider content. Besides cross-sectional research, with representative cohorts and samples, longitudinal studies might prove useful.

For reasons of convenience, the studies we carried out involved university students. These are often heavy users of the technologies under examination, as national surveys and other data show. It was intriguing to monitor the development of the representation in a critical phase of development of the new devices. However, it would be worthwhile broadening the inquiry to take into account individuals who relate to ICTs from different positions and roles, as well as those of different generations. We would thus make operational the importance of memberships of different groups, practices, and values at the theoretical and empirical level in the outline of the social systems of knowledge.

A wide-ranging inquiry on the theme of social thinking related to ICTs, however, should not restrict itself to the analysis of free associations and might benefit from the study of retrospective interviews, as in, for example, the episodic interviews accomplished by Flick (1996) in order to study the social representations of the technological change in IT engineers, academics and teachers' daily life, in France and in East and West Germany in the late 1980s. Moreover, investigations of the mass media (mainly newspapers) would be valuable, aiming to isolate the various features—according to Flick, the 'bricks' which constitute the texture—and the underlying dimensions, the holding structures, and the organizing principles which take form in the ongoing play between media communication and interpersonal communication in everyday life.

Notes

[1] To measure attitudes, we turned to free answers and to a semantic differential scale. In the questionnaire, we included further stimuli and scales and we collected information on practices relative to the mobile and on the affective and relational network of the participants. However, as these are not the focus of the present study, they will not be taken into account here.

[2] The positioning of an illustrative variable on a dimension is measured by means of a *valeurs-test* (v–t). When a v-t value is greater than 2 in absolute value, the chances that this positioning is not due to hazard is 95 per cent. Only modalities with a v-t equal to or greater than |2| are presented in the paper.

[3] The absolute contribution of a point to a dimension is the proportion of inertia (variance) explained by the point. The sum of the contributions of the points to each factor is equal to 100. Only points with considerable value (i.e. $>100/n$, where n is the number of points) are discussed.

[4] This amount of inertia, which would be considered wholly insufficient in factor analysis, in the case of correspondence analysis on a cases × participants matrix is far from unusual and allows to draw firm conclusions in terms of co-occurrences of the considered items (cf. Lebart & Salem, 1988).

References

Benzécri, J. P. *et al.* (1973) *L'Analyse de données*, vol. 2: *L'Analyse des correspondances*, Dunod, Paris.

Brown, B., Green, N. & Harper, R. (eds) (2002) *Wireless World: Social, Cultural and Interactional Issues in Mobile Communications and Computing*, Springer, London.

Capozza, D., Falvo, R., Robusto, E. & Orlando, A. (2003) 'Beliefs about Internet: methods of elicitation and measurement', *Papers on Social Representations*, vol. 12, pp. 1–14.

Contarello, A. & Fortunati, L. (2006) 'ICTs and the human body: a social representation approach', in *New Technologies in Global Societies*, eds P. Law, L. Fortunati & S. Yang, World Scientific, Singapore.

Contarello, A., Fortunati, L., Gomez, P., Mante, E., Vershinskaya, O. & Volovici, D. (2005) 'General trends in digital culture', in *UNESCO between Two Phases of the World Summit on the Information Society*, Proceedings of the international conference held in Saint Petersburg, Russian Federation, 17–19 May, Institute of the Information Society, pp. 123–132.

Contarello, A. & Sarrica, M. (2007) 'ICTs, social thinking and subjective well-being: the Internet and its representations in everyday life', *Computers in Human Behavior*, vol. 23, no. 2, pp. 1016–1032.

Di Giacomo, J. P. (1980) 'Intergroup alliances and rejections within a protest movement: analysis of the social representations', *European Journal of Social Psychology*, vol. 10, pp. 329–344.

Doise, W., Clémence, A. & Lorenzi-Cioldi, F. (1993) *The Quantitative Analysis of Social Representations*, Harvester Wheatsheaf, Hemel Hempstead.

Farr, R. & Moscovici, S. (eds) (1984) *Social Representations*, Cambridge University Press, Cambridge.

Flick, U. (1996) *Psychologie des Technisierten Alltags*, Westdeucher, Opladen.

Fortunati, L. (1998a) 'The ambiguous image of the mobile phone', in *Communications on the Move: the Experience of Mobile Telephony in the 1990s*, ed. L. Haddon, Telia, Stockholm.

Fortunati, L. (ed.) (1998b) *Telecomunicando in Europa*, Angeli, Milan.

Fortunati, L. (2001) 'The mobile phone: an identity on the move', *Personal and Ubiquitous Computing*, vol. 5, no. 2, pp. 85–98.

Fortunati, L. (2005) 'Mobile phones and fashion in post-modernity', *Telektronikk*, vol. 3, no. 4, pp. 35–48, Available also at: http://www.telenor.com/telektronikk/volumes/pdf/3_4.2005/Page_035-048.pdf (last accessed 23 Mar. 2007).

Fortunati, L. & Contarello, A. (2002) 'Internet–mobile convergence: via similarity or complementarity?', *Trends in Communication*, vol. 9, pp. 81–97.

Fortunati, L. & Contarello, A. (2005) 'Social representation of the mobile: an Italian study', in *When Mobiles Came*, ed. S. D. Kim, Communication Books, Seoul.

Fortunati, L., Katz, J. E. & Riccini, R. (eds) (2003) *Mediating the Human Body*, LEA, Mahwah, NJ.

Fortunati, L. & Manganelli, A. M. (2007) 'The social representation of telecommunications', *Personal and Ubiquitous Computing*, DOI: 10.1007/s00779-006-0139-7 (last accessed 26 Mar. 2007).

Gergen, K. J. (2002) 'The challenge of absent presence', in *Perpetual Contact: Mobile Communication, Private Talk, Public Performance*, eds J. E. Katz & M. Aakhus, Cambridge University Press, Cambridge.
Gergen, K. J. (2003) 'Self and community in the new floating worlds', in *Mobile Democracy: Essays on Society, Self and Politics*, ed. K. Nyíri, Passagen, Vienna.
Geser, H. (2004) 'Towards a sociological theory of the mobile phone', Available at: http://socio.ch/mobile/t_geser1.pdf (last accessed 26 Mar. 2007).
Glotz, P., Berschi, S. & Locke, C. (eds) (2005) *Thumb Culture: the Meaning of Mobile Phones for Society*, Transcript, Bielefeld.
Green, N., Harper, R. H. R., Murtagh, C. & Cooper, G. (2001) 'Configuring the mobile user: sociological and industry views', *Personal and Ubiquitous Computing*, vol. 2, pp. 146–156.
Haddon, L. (ed.) (1998) *Communications on the Move: the Experience of Mobile Telephony in the 1990s*, Telia, Stockholm.
Haddon, L., Mante, E., Sapio, B., Kommonen, K. -H., Fortunati, L. & Kant, A. (eds) (2005) *Everyday Innovators: Researching the Role of Users in Shaping ICTs*, Springer, Dordrecht.
Höflich, J. R. & Gebhaedt, J. (eds) (2005) *Mobile Kommunikation: Perspektiven und Forschungsfelder*, Peter Lang, Frankfurt.
Katz, J. (ed.) (2003) *Machines that Become Us*, Transaction, New Brunswick, NJ.
Katz, J. & Aakhus, M. (eds) (2002) *Perpetual Contact: Mobile Communication, Private Talk, Public Performance*, Cambridge University Press, Cambridge.
Kim, D. S. (ed.) (2005) *When Mobile Came: The Cultural and Social Impact of Mobile Communication*, Communication Books, Seoul.
Law, P., Fortunati, L. & Yang, S. (eds) (2007) *New Technologies in Global Societies*, World Scientific, Singapore.
Le Bouedec, G. (1984) 'Contribution à la méthodologie d'étude des représentations sociales', *Cahiers de Psychologie Cognitive*, vol. 4, pp. 245–272.
Lebart, L., Morineau, A., Bécue, M. & Haeusler, L. (1989) *Système portable pour l'analyse des données textuelles (Spad.T)*, Cisia, Paris.
Lebart, L. & Salem, A. (1988) *Analyse statistique des données textuelles*, Dunod, Paris.
Licoppe, C. & Heurtin, J. P. (2001) 'Managing one's availability to telephone communication through mobile phones: a French case study of the development dynamics of mobile phone use', *Personal and Ubiquitous Computing*, vol. 5, no. 2, pp. 99–108.
Ling, R. (ed.) (2001a) 'Mobile communication and the reformulation of the social order', *Personal and Ubiquitous Computing*, vol. 5, no. 2, pp. 83–84.
Ling, R. (2001b) '"We release them little by little": maturation and gender identity as seen in the use of mobile telephony', *Personal and Ubiquitous Computing*, vol. 5, no. 2, pp. 123–136.
Ling, R. & Pedersen, E. (eds) (2005) *Mobile Communication and the Re-negotiation of the Social Sphere*, Springer, London.
Lorente, S. (ed.) (2002) 'Juventud y teléfonos móviles', (special issue), *Revista de Estudios de Juventud*, vol. 57.
Meyrowitz, J. (2003) 'Global nomads in the digital veldt', in *Mobile Democracy: Essays on Society, Self and Politics*, ed. K. Nyíri, Passagen, Vienna.
Meyrowitz, J. (2005) 'The rise of glocality: new senses of place and identity in the global village', in *A Sense of Place: the Global and the Local in Mobile Communication*, ed. K. Nyíri,, pp. 21–30.
Moliner, P. (1993) 'Cinq questions à propos des représentation sociales', *Cahiers Internationaux de Psychologie Sociale*, vol. 20, pp. 5–14.
Moscovici, S. (1961/76) *La Psychanalyse: son image et son public*, PUF, Paris.
Moscovici, S. (1998) 'The history and actuality of social representations', in *The Psychology of the Social*, ed. U. Flick, Cambridge University Press, Cambridge.
Nyíri, K. (ed.) (2003) *Mobile Democracy: Essays on Society, Self and Politics*, Passagen, Vienna.

Nyíri, K. (ed.) (2005) *A Sense of Place: the Global and the Local in Mobile Communication*, Passagen, Vienna.

Palen, L., Salzman, M. & Youngs, E. (2001) 'Discovery and integration of mobile communications in everyday life', *Personal and Ubiquitous Computing*, vol. 5, no. 2, pp. 109–122.

Rheingold, H. (2002) *Smart Mobs: the Next Social Revolution*, Perseus, New York.

Silverstone, R. & Hirsch, E. (1992) *Consuming Technologies: Media and Information in Domestic Spaces*, Routledge, London.

Wagner, W. & Hayes, N. (2005) *Everybody Discourse and Common-Sense. The Theory of Social Representations*, Palgrave-Macmillan, Basingstoke.

Wagner, W., Kronberger, N. & Seifert, F. (2002) 'Collective symbolic coping with new technology: knowledge, images and public discourse', *British Journal of Social Psychology*, vol. 41, pp. 323–343.

Weilenmann, A. (2001) 'Negotiating use: making sense of mobile technology', *Personal and Ubiquitous Computing*, vol. 5, no. 2, pp. 109–122.

Illusions of Balance and Control in an Always-on Environment: a Case Study of BlackBerry Users

Catherine A. Middleton

Introduction

This paper presents a qualitative case study of Canadian BlackBerry® users. It begins with a brief description of the BlackBerry, a handheld wireless mobile e-mail device developed by Research in Motion.[1] BlackBerry users find their devices to be empowering, allowing them more control over their environments. The BlackBerry does give its users a mechanism to exert control over the management of daily communication tasks, but by virtue of its always-on, always-connected nature, it also reinforces cultures that expect people to be accessible outside normal business hours. Rather than just a tool of liberation for its users, the BlackBerry can also be understood as an artefact that reflects and perpetuates organizational cultures in which individual employees have little control and influence. While this case study focuses on BlackBerry users, it is suggested that the findings are not unique to this device. BlackBerries and other mobile technologies have been envisioned by some as means of enforcing work–life boundaries, but this paper concludes that the use of always-on mobile devices can lead to situations where conflict between work and personal activities is exacerbated rather than reduced.

Since its launch in 1999, the BlackBerry has established itself as the mobile e-mail device of choice for business and government sectors.[2] Although it offers a mobile phone, calendaring, contact and task management, and Web browsing, the main reason for having a BlackBerry is to send and receive e-mails. The device has been very successful in providing easy-to-use, always-on, reliable e-mail connectivity from virtually anywhere.[3]

Unlike competing devices, the BlackBerry embodies a distinct cultural identity, and has become part of the 'cultural universe' (Du Gay *et al.*, 1997) in North America and beyond.[4] The BlackBerry has been the subject of much attention in the popular press.[5] Stories about BlackBerry usage are often predictable, offering anecdotes of users who cannot be separated from their devices, discussions of the addictive nature of the devices (frequently invoking the word 'Crackberry'), and identification of aggrieved spouses, partners or colleagues who resent the BlackBerry's intrusiveness (e.g. Economist Staff, 2005; Parks, 2006; Terry, 2006).

Curiosity about such extreme reactions to a simple mobile device prompted a study of Canadian BlackBerry users. Initial analysis showed how mobile e-mail usage can be both dysfunctional and functional, arguing that assessment of mobile e-mail is influenced by organizational culture (Middleton & Cukier, 2006). In this paper, Schein's (2004) model of organizational culture is applied as a means of further exploring the cultural values and assumptions that influence patterns of BlackBerry use. The cultural framework can help to reconcile the disparate perceptions of the BlackBerry by showing how these perceptions are embedded in different layers of cultural understanding. When examined as a cultural artefact, the BlackBerry is a tool of efficiency, providing control over users' communication needs from any location, and enabling responsiveness and accessibility at all times. Examining the values that encourage these usage patterns allows identification of work intensification and work extension practices. In turn, these routines reflect organizational cultures in which it is assumed that employees are available in times and places extending well beyond office hours. The paper concludes by discussing the culture of BlackBerry usage within the context of work–life balance research, suggesting that the adoption of mobile technologies may lead to increased work–life conflict, resulting in detrimental organizational outcomes.

Literature Review

There is an extensive body of work discussing consumer usage of mobile technologies, with a particular focus on mobile telephony (see, for example, Brown *et al.*, 2002; Fortunati, 2002; Gordon, 2002; Haddon, 1997; Humphreys, 2005; Katz, 2003; Katz & Aakhus, 2002; Ling, 2004; Nyíri, 2003; Oksman & Turtiainen, 2004; Peters & Allouch, 2005). In contrast, there is limited academic work on organizational usage of mobile e-mail. In the past few years, BlackBerries have become commonplace among business users, and their attributes have become known to the general public (especially as a patent dispute threatened to shut down the service; see ' BlackBerry blackout', 2005), but the devices have received limited notice in studies of technology usage by mobile professionals. Nah *et al.* (2005) and Sørensen and Gibson (2004) do note the BlackBerry's perceived value.

Allen & Shoard (2005) investigated whether BlackBerries would be effective in managing e-mail overload. Their results indicate that adoption of BlackBerries did not reduce perceived overload, but did allow users (senior police officers) greater control over their e-mail because they could read and respond to it throughout the day, regardless of whether or not they were in the office. Mazmanian *et al.* (2005) report that mobile e-mail users were expected to, and did, respond more quickly to e-mail than people who did not have access to a mobile device.[6] Users reported 'constant and sustained interaction with their devices' (p. 340), using them to monitor their work environments. The users valued the increased connectivity that the devices enabled, but reported spousal resentment toward the intrusions that such connections created. Jarvenpaa et al.'s (2005) work notes user ambivalence to mobile technologies in a series of paradoxes. For example, mobile technologies both empower and enslave users, simultaneously engage and disengage them, and blur the boundaries between private and public space and communication.

For their part, Middleton and Cukier's (2006) initial analysis of mobile e-mail has shown how it could be constructed as functional and dysfunctional. For instance, in the name of efficiency (a functional attribute) mobile e-mail users carried out dangerous practices (such as e-mailing while driving, an example of a dysfunctional use of mobile e-mail). Minimally disruptive behaviours were also considered anti-social, the immediacy of mobile e-mail could be distracting, and by liberating users from the confines of work, mobile e-mail delivered work to the home and to other non-work environments, challenging work–life boundaries. Cousins & Robey (2005) address the issue of boundary control in their study of nomadic work patterns, concluding that the workers they studied used mobile technologies in ways that sharpened and reinforced work–life boundaries. But while these respondents may have believed that they were keeping work time and personal time separate by defining clear boundaries, it seems that work time was dominant for at least some of the respondents. As such, boundaries may have been maintained, but it is not clear that such boundaries resulted in balance between work and personal activities. Govindaraju & Sward (2005) also argued that mobile technologies can enable work–life balance. By 'time slicing' (using very small portions of time to do work activities), people could be productive even when they only had a few minutes to spare. Time was also 'shifted', allowing people to move work around during the day to accommodate personal activities. The researchers expressed concern that the availability of mobile technologies that increased people's ability to work anywhere would be perceived as encroaching on personal time, but their respondents (Intel employees) felt that the increased flexibility was beneficial.

There are advantages (mobility, remote control over work) and disadvantages (accessibility, blurred work–life boundaries) that accrue to users as they adopt mobile technologies. This paper offers further investigation into the relationship between mobile technologies and work–life balance, an important issue as the adoption of always-on devices like BlackBerries becomes more prevalent in the workplace.

Research Methodology

Data presented here come from semi-structured, qualitative interviews conducted with 13 Canadian BlackBerry users, as part of a larger study exploring contradictions of mobility (described in Middleton *et al.*, 2005). Respondents were identified by the author and a colleague through their networks of business contacts, with those participating in the study representing a variety of industries, differing levels of job seniority, and ranging in age from their mid-20s to their mid-50s. Respondents were not primarily 'nomadic' or 'mobile' workers, but frequently moved around the cities in which they lived (Montreal, Ottawa, Toronto, Hamilton, Regina). Most of their jobs involved some travel. A limitation in the data set is the gender imbalance. Of the 13 interviews discussed in this paper, only two were with female BlackBerry users. All interviews were audiotaped and professionally transcribed. Thematic categories were established from the initial exploratory research questions, and coding was done using the ATLAS.ti qualitative data analysis software package. This paper focuses on data describing the BlackBerry as a cultural artefact, and data that outline the routines established in using the device.

The BlackBerry Experience

The analysis adopts Schein's model of organizational culture (Schein, 2004) as a means of understanding BlackBerry use within organizations. Schein's model shows how cultures are established and perpetuated within organizations, through an exploration of artefacts, values and underlying assumptions.[7].

Respondents in this study come from diverse backgrounds, and at a macro level experience varied organizational cultures. In applying a cultural analysis, it is not suggested that respondents' cultures are uniform in all aspects. It is argued, however, based on initial analysis of these data (Middleton & Cukier, 2006), that there is a commonality of cultural influences impacting the usage of mobile communication devices within respondents' organizations and work environments. This commonality is demonstrated in the thematic congruence observed in participants' experiences with their devices and reflects shared understandings of the value of mobile devices in their work environments.

Schein (2004, p. 8) argues that culture 'guides and constrains' behaviours within groups. Artefacts (including technologies, rituals, myths and stories) are visible manifestations of culture, 'easy to observe, and very difficult to decipher' (p. 26). To gain a deeper understanding of the way an artefact is used and understood, a study of organizational values relevant to the artefact is needed. Values reflect guiding social principles (Hatch, 1993) and can be understood by examining organizational practices and activities (Van Den Berg & Wilderom, 2004). Underpinning organizational values are the 'basic assumptions' that encompass 'taken-for-granted' shared understandings on which actions are premised. These basic assumptions enable the development of shared 'mental maps' that guide behaviours among group members (Schein, 2004, pp. 30–32).

The BlackBerry in Action

It is difficult to understand the level of engagement that BlackBerry users have with their devices without witnessing users in action. This section offers a brief composite view of a typical BlackBerry user, drawn from user data and popular press accounts. The first thing many users do in the morning is check their BlackBerry, and the device remains with them throughout the day, until they go to bed. Some users even admit to waking up to read e-mail in the middle of the night (Maiden, 2006). An anonymous poster to the techdirt.com website describes the BlackBerry as an 'Electronic Pet'—'you constantly see people taking their little pets out and stroking the scroll wheel, coddling them, basically "petting" them' (Anonymous, 2004). It is common for BlackBerry users to respond in a Pavlovian fashion to the beeps or buzzes that indicate the arrival of messages, grabbing at the device immediately to see what is new. In organizations with multiple BlackBerry users, notification and clicking sounds fade into the background as a familiar backdrop to business activities. BlackBerries 'turn e-mail from something you do on a computer to something you just kind of do ... you go from something that is done at a desk to something that is done wherever you are', many, many times a day. Demonstrating that BlackBerries do go everywhere, one respondent damaged his BlackBerry when he was caught in a wave on a beach, another damaged hers when it fell into a toilet. The BlackBerry is described as a cultural artefact below, and user routines are discussed in more detail when considering the values that underpin BlackBerry usage.

The Artefact

Users' descriptions of the BlackBerry and its functionality were straightforward, focusing on its ease of use, mobility, flexibility, convenience, always-on immediacy, and reliability. As one user said: 'it simply does what needs to be done'. Despite the BlackBerry's reputation as an addictive device, the informants did not cast their devices in that light. The word 'Crackberry' was mentioned in only three interviews, twice in reference to someone else's description of the device. While several users did acknowledge the addictive qualities of their devices, it was far more common for users to emphasize and reinforce the BlackBerry's benefits. Descriptors included liberating, freeing, tremendously useful and beneficial.

Values

The terms used to describe the artefact offer insights into respondents' values about mobility. Although individuals worked for different organizations, their values were similar and led to common behaviours. Values revolved around efficiency, immediacy, accessibility, and flexibility, and were realized in the routines that users adopted with their BlackBerries.

Work intensification

Many of the respondents described using their BlackBerries in ways that intensified their working practices, so that they could accomplish additional tasks within the traditional time and space confines of their jobs. For example, the device was used as a means of silent communication when attending a meeting, allowing the user to reach out to the external world beyond the meeting. Indeed, once colleagues were aware that someone had a BlackBerry, being in a meeting was no longer an excuse for not responding to a message immediately.

For many respondents, the BlackBerry offered a sense of security, by allowing them to monitor situations remotely. People spoke of the ability to deal with urgent situations or 'dilemmas' by e-mail, rather than by disrupting meetings by 'jumping out and taking the phone call'. 'Having a quick look' at the device when a new message was received provided reassurance, even if the user did not choose to open the message or reply to it. One user described the BlackBerry as being like a car 'dashboard', allowing for a quick scan of various items. Another described it as a means of 'directing traffic', an essential activity as he juggled the multiple demands of his job.

Although they tried to be discreet when using their BlackBerries, users were very aware of the disruptive nature of these intensified work practices. When asked about attending meetings where people did use their BlackBerries throughout, there was an air of resignation indicating that while this wasn't the optimal practice, it did allow people to manage the multiple demands on their time. One workplace had instituted a 'no BlackBerry' rule for meetings, but in others respondents rationalized their own and others' BlackBerry usage in meetings by suggesting that given the alternative of skipping the meeting, 'occasionally checking their Blackberry' was acceptable.

Time-based work extension

In addition to making it possible to intensify work practices, the BlackBerry made it much easier for individuals to work longer hours. While many of these individuals worked long hours before getting their BlackBerries, certain affordances of the device were particularly conducive to extending work hours. For instance, because the device has an alarm clock, several respondents reported that the last thing they did at night and the first thing they did in the morning was check their e-mail, as the BlackBerry was at their bedside. To the dismay of partners and spouses, e-mail monitoring behaviours continued throughout most respondents' waking hours. One user noted 'if I've misplaced my BlackBerry there's an immediate understanding around the home that I'm definitely going to be on edge until it's found'. Although some people did turn off their BlackBerries, or at least 'take them off their belts' when they got home, others kept the device active and in reach at all times. Only one organization explicitly freed people from their devices by instituting a policy of 'no e-mail' after 6.00 p.m. or on weekends. In contrast, others reported having 'conversations all the time with my boss at ten or eleven o'clock at night, with the BlackBerry', and 'getting Berries at all points, weekends, eight o'clock this morning and probably eight o'clock tonight'.

Location-based work extension

Mobile devices are valued by many because they offer anytime, anywhere availability. In most places users simply turn on the device and e-mail appears in the inbox. Some users were frustrated that their service provider did not offer international connectivity, but for those who did have this option, using the BlackBerry outside the country was seamless. Individuals used their BlackBerries on aircraft (turning on the transmission signal after the plane landed to send out queued messages instantly), trains, buses, and in cars and taxis. Two respondents 'berried' on their bicycles, and one person commented on almost being hit by BlackBerry-wielding pedestrians not looking where they were going. (This scenario is the basis of a comedy sketch, promoting the 'BlackBerry Helmet'; see Rick Mercer Report, 2006.).

With a BlackBerry, e-mail 'follows' the user around. Many chose to have their e-mail follow them even while on vacation, although some were 'instructed' by their spouses to leave the device behind. Commenting about using his device on the golf course, one respondent noted that 'maybe you do more work, but it frees you, it just lets you get out of the office without disrupting your ability to stay connected'. Another, one of three who reported using the BlackBerry at the beach said, 'Look how empowering BlackBerry is. I can be on the beach with the family and be connected in. That is a great comfort to me.'

For these users, having a device that connected them to the office and the business environment from any location was empowering, not intrusive. That the device was on all the time was not seen as an infringement on personal time, but a means of controlling the work environment to better fit personal needs, offering liberation, freedom and peace of mind.

These routines reflect the values of BlackBerry users, and by extension offer insights into their organizations' values around mobile technologies. The routines that enabled users to intensify their work practices using mobile devices, and to work anytime and anywhere, reflect organizational cultures that value immediacy and responsiveness. No respondents spoke directly of organizational mandates requiring 'always-on' availability, but it was clear from their reflections about BlackBerry usage that being connected to their jobs well beyond the traditional 'nine to five' business hours was very important to them.

Underlying Assumptions

The BlackBerry users who participated in this survey did not question their commitment to long working hours and were totally engaged with their jobs. In the organization that discouraged evening and weekend e-mail, employees still worked at those times. Describing his job, one respondent said, 'I love the work I do, I would be available 24 hours.'[8] Users accepted the fact that they were on call for up to 24 hours a day, and embraced the challenges and responsibilities of their jobs. Users got BlackBerries to help cope with the demands of their jobs.

The unspoken assumption among respondents was that the environments in which they worked created pressures to be engaged and involved with their jobs. The BlackBerry is an instrument of engagement with this environment. Using the terminology presented in a large-scale study of Canadian workers, our respondents' work environments could be characterized as having a 'culture of hours' (Duxbury *et al.*, 2003). In this type of culture, it is understood that long working hours are necessary for success and advancement (Higgins & Duxbury, 2005). The cultures experienced by the BlackBerry users could further be described as 'cultures of delivery' or 'cultures of performance', where the focus is on doing 'whatever it takes' to get the job done.

Reflecting on BlackBerry Users' Cultures of Mobility

This paper uses a cultural lens to understand how the usage of a specific mobile device becomes routinized and embedded within organizations. A discussion of the BlackBerry as a cultural artefact shows the construction of the BlackBerry as a tool of efficiency. This understanding is reinforced by observed routines of usage, routines that take advantage of the device's mobility and discreet nature, combined with the power of e-mail, to encourage work intensification, and the extension of work into times and spaces more traditionally considered non-work or personal times and spaces. The routines reflect organizational values of availability and responsiveness, which in turn reflect deep-seated organizational beliefs about employee commitment to long working hours, and to the achievement of 'results'. The tool of efficiency is also an instrument that connects employees to their organizations, facilitating and possibly encouraging long working hours.

The interviews with BlackBerry users demonstrate their commitment to this vision of their work worlds. Although there are, of course, some points of resistance to this perspective, a persistent theme focused on the value of the BlackBerry in allowing its users to cope with the extended demands on their time and goodwill that are accepted as 'the nature of the job' in today's (North American) business environment. Users did acknowledge that using their BlackBerries could create intrusions, but always rationalized their usage in ways that demonstrated how the device could help to manage such intrusions.

These BlackBerry users resisted the 'Crackberry' label so frequently used by the media, and did not generally consider their usage of the device to be addictive. If they were to admit an addiction, it would likely be to their work, not to a tool they used to manage their work. Nevertheless, the tool facilitated their work addiction, by seamlessly bringing the office to the worker on a 24/7 basis.

If this paper were to conclude here, it would overlook some key aspects of BlackBerry usage by individuals in organizations. Thus far, the paper has presented detailed insights into BlackBerry usage, refuting the Crackberry claims of addiction (claims made primarily by non-users), and explaining how users feel that the device offers liberation and control over their work environments. This line of argument is supported by the data, and reflects a consistency in users' organizational cultures at the levels of the mobile artefact, values and underlying assumptions about mobility.

But the Crackberry idea should not be so easily dismissed. Whether or not users acknowledge it, there is a case to be made that using a BlackBerry (or any other mobile device that enables the same type of seamless 24/7, work–life integration) supports and perpetuates the very work culture that users are trying to manage and control through the device. This is an instance of Metcalf's law (Gilder, 1993), which in simple terms suggests that communication networks become more valuable the more users join them. Mobile device usage begets more mobile device usage, addictive or not. The more that individuals make themselves electronically open and available to their colleagues, clients and others, the more this availability will be exploited. Actions that appear as reasonable attempts to control a demanding job can encourage further engagement, resulting in increased, rather than decreased, workload.

While it may be too strong to suggest that a crisis is looming, mobile device adoption in the workplace does facilitate situations in which work–life balance is compromised. As noted above, in and of themselves, neither the BlackBerry nor similar mobile devices further erode an already tenuous work–life balance. What is different with mobile devices as compared to older telecommuting, home-based technologies is that the mobile devices are more efficient at engaging individuals in work activities. Mobile devices have been embraced because they allow people to work anywhere and at any time, but such affordances have the potential to increase work–life conflict.

Duxbury, Higgins and colleagues have been investigating balance and conflict among work, family and life for many years (Duxbury & Higgins, 2001; Duxbury *et al.*, 1994, 2003). Higgins & Duxbury (2005) reflect on this research, noting that in multiple studies since 1989, each of three elements of work–life conflict has been reported more frequently over time. It is easy to make links between these elements and the data presented above.

'Role overload' is experienced when a person has 'too much to do in the available time' (Higgins & Duxbury, 2005, p. 1). This situation was faced regularly by the BlackBerry users in this study, and one response to it was to use the BlackBerry to adopt work intensification behaviours. With the BlackBerry, users tried to do more than one thing at once, so that they could cram more into their tight schedules. Work intensification was observed within the confines of their existing daily routines, and was carried out in addition to observed work extension routines. Hallowell (2005) challenges the effectiveness of work intensification practices, arguing that multitasking of this nature is not successful.

Work extension practices were not new to the BlackBerry users, many of whom worked long hours from multiple locations before they obtained their BlackBerries. What was different about adopting the mobile device was that it made it easier to further extend work into personal time and space. For example, having the BlackBerry beside the bed not only extended work but also created work–family interference. Another source of work–family interference was noted when users took their mobile devices to places where earlier technologies did not function well. For example, the small size and long battery life made it possible to use BlackBerries while camping, on remote beaches, on swimming pool decks and at vacation homes.

It is very easy to toss a BlackBerry into a purse or clip it onto a belt, meaning that the user is rarely without it. For this reason, in addition to work–family interference, BlackBerry users also experienced family–work interference. As one young father explained: 'work follows you home, home follows you to work', adding that 'for me that's been a good thing'. Many users commented on instances where their device brought a family or personal issue into their work environment, and this was not perceived as being disruptive. Among this particular group of users there were fewer instances of family–work overlap than of work–family overlap, which is perhaps a reflection of the gender imbalance among the respondents. (For example, Chesley, 2005, found that mobile phone usage was associated with increased negative family–work 'spillover', or imbalance, in women, but not in men, although phones were associated with increased negative work–family spillover for both genders.).

These data on BlackBerry usage fit well with research on work–life conflict. Work intensification practices can be understood as a response to role overload. Work extension practices put stress on work–family relationships, increasing the potential for conflict as mobile devices became an unwelcome part of their users' personal lives (unwelcome at least to those around them). Family–work interactions were supported by the mobile devices, and did not appear to be problematic.

Duxbury *et al.*, (1994) explain that work–life stress is mitigated by perceived control. This relationship provides a plausible explanation as to why the respondents constantly rationalized as acceptable what appeared to be intrusive uses of their devices. An outside observer could easily conclude that behaviours like reading e-mail at midnight or checking in with the office before getting out of bed in the morning are indicative of work–life conflict. But the source of conflict is in the work, and the device is used in an attempt to control the demands of the job. Nevertheless, it is difficult to imagine that even if a person is choosing to work at midnight and first thing in the morning that such behaviours reflect a balanced work–life environment.

Looking at BlackBerry usage in the context of work–life balance helps to provide perspective on the behaviours reported here. For most users, BlackBerries are not helping in achieving work–life balance. Despite the sense of control enabled by BlackBerries, their always-on nature encourages behaviours that make it easy to do more work and spend less time pursuing personal and family activities.

It is beyond the scope of this paper to address the implications of decreasing work–life balance. This topic is the subject of an extensive report by Duxbury & Higgins (2001), in which they clearly outline the negative societal consequences of work–life conflict and offer remedies for employers, employees and government to begin addressing the situation.

Conclusions

The paper started with the observation that BlackBerries are loved by their users yet frequently loathed by their users' closest friends and relations. Data were presented to show why BlackBerries were so highly regarded by their users, demonstrating their versatility in supporting anywhere, anytime work practices, and enabling users to be more efficient and to do more work throughout their busy days. Users explained that their BlackBerries are tools to help them control and manage their work environments. They acknowledged working long hours, but loved their jobs and valued the flexibility that their BlackBerries offered to manage their work on their own terms. The data offer a compelling argument as to why BlackBerries have become so prevalent in business and government circles in North America.

But the paper also argues that this situation is not as positive as it might appear at first glance. Without directly challenging the opinions of the people who participated in this study, it is suggested that their perceptions of acceptable engagement levels with their mobile devices are influenced by organizational cultures that reinforce overwork and promote unrealistic expectations for employee engagement in their jobs. The respondents' beliefs that justify and rationalize their usage of their BlackBerries at all times of day and night, from all over the world, whether working or on vacation, are sincere and appear to be deeply internalized. But this paper provides an interpretation of these beliefs that shows how they are shaped by arguably unreasonable organizational demands and expectations regarding employee commitment and accessibility.

The paper considers these findings from the perspective of the work–life balance, observing that work extension and intensification behaviours are likely to increase work–life conflict. It is suggested that, over time, the widespread adoption of mobile technologies may result in detrimental organizational outcomes, unless organizations are better able to manage employee workloads and set reasonable expectations for engagement with mobile devices.

A potentially interesting line of inquiry for future research in this area would be to investigate mobile device usage through an organizational learning perspective. The question of interest here would be how organizations and the individuals in them learn to cope with what are likely to be the increasingly negative impacts of mobile technology adoption in the workplace. Can organizations practise 'double loop learning'[9] (Argyris & Schön, 1978) to reassess and adjust their courses of action? Are organizations trapped in feedback loops that reinforce the negative aspects of always-on connectivity, or can they 'learn' to change (Senge, 1990)? Do organizations face 'escalating commitment' to courses of action (Staw, 1981), and, if so, can they resolve such situations appropriately?.

In closing, it is reiterated that the issues discussed here are not just BlackBerry issues. Problems of overwork and imbalance between personal and work commitments have been a part of (some) organizations for years. What has changed, though, is the ease with which work can extend so deeply into employees' personal time and spaces. BlackBerries have been successful because they do allow people to work anywhere, anytime. But because they allow people to work anywhere, anytime they reinforce negative aspects of organizational cultures that encourage overwork.

Notes

[1] 'BlackBerry' is a registered trademark of Research in Motion (RIM).

[2] Refer to www.rim.com/news/press/index.shtml for quarterly updates of subscriber numbers. As of June 2006 subscriber numbers were estimated at approximately 5.5 million. It is estimated that RIM has 75 per cent of the US mobile e-mail market (Waldie, 2006).

[3] One of the appeals of the BlackBerry is its seamless international roaming ability. Users must pick a provider that supports this service. With this service, simply turning on the device starts the flow of e-mail, from any worldwide location within network reach.

[4] It is noted that the BlackBerry is most widely adopted and best known in the United States and Canada (Reguly, 2006). In the past couple of years it has gained a foothold among mobile users in the United Kingdom (Tait, 2006) and Australia (Connors, 2006). At the time of writing, the BlackBerry was not widely used in European or Asian markets, where 3G mobile phones often provide more functionality than BlackBerries. Service has been available in India since 2004 (Sharma, 2004). There is limited BlackBerry adoption in South America, with service becoming available in 13 countries in 2006 (BlackBerry Press Releases, 2005). A full list of products available around the world can be found online at http://www.rim.net/products/handhelds/index.shtml.

[5] Examples can be found by searching databases including ABI/Inform, Factiva or LexisNexis, using search terms like 'BlackBerry and Crackberry'.

[6] It is assumed from the use of the word 'Crackberry' in the article title that the 'wireless email devices' studied were BlackBerries, but this is not stated explicitly.

[7] Hatch (1993) provides a synthesis of research criticizing Schein's approach to organizational culture. She proposes a reformulation of his model incorporating symbols, and focusing on the dynamics of culture, but notes that Schein's model 'continues to have relevance' (p. 658). Cultural analysis is used in this paper as a diagnostic tool to expose shared understandings and underlying assumptions embedded in the adoption of an artefact. Schein's model is well suited for this purpose.

[8] An interesting extension of work on BlackBerry usage would be to consider the personalities of individual adopters, with a specific focus on traits that indicated workaholism. Work by Burke and colleagues (Burke, 2001, 2004; Burke *et al.*, 2004) and Buelens & Poelmans (2004) could provide starting points for this line of investigation.

[9] Double loop learning involves questioning the norms in which an organization operates. A double loop learning response may involve challenging organizational norms, whereas a single loop process operates within existing norms (Morgan, 1997, ch. 4).

References

Allen, D. K. & Shoard, M. (2005) 'Spreading the load: mobile information and communications technologies and their effect on information overload', *Information Research*, vol. 10, no. 2, Paper 227, n.p.

Anonymous (2004) 'The life of a BlackBerry widow', tech dirt, Available at: http://www.techdirt.com/articles/20040503/0224203_F.shtml (accessed 27 Mar. 2007).

Argyris, C. A. & Schön, D. A. (1978) *Organizational Learning: a Theory of Action Perspective*, Addison-Wesley, Reading, MA.

'BlackBerry blackout' (2005) *Wall Street Journal*, 30 Nov., p. A18.

BlackBerry Press Releases (2005) 'Telefónica Móviles and Research in Motion to launch BlackBerry across Latin America', Available at: http://www.blackberry.com/news/press/2005/pr-30_11_2005-02.shtml (accessed 27 Mar. 2007).

Brown, B., Green, N. & Harper, R. (eds) (2002) *Wireless World: Social and Interactional Aspects of the Mobile Age*, Springer, London.

Buelens, M. & Poelmans, S. A. Y. (2004) 'Enriching the Spence and Robbins typology of workaholism: demographic, motivational and organizational correlates', *Journal of Organizational Change Management*, vol. 17, no. 5, pp. 440–458.

Burke, R. J. (2001) 'Workaholism in organizations: the role of organizational values', *Personnel Review*, vol. 30, no. 5/6, pp. 637–645.

Burke, R. J. (2004) 'Introduction: workaholism in organizations', *Journal of Organizational Change Management*, vol. 17, no. 5, pp. 420–423.

Burke, R. J., Richardsen, A. M. & Martinussen, M. (2004) 'Workaholism among Norwegian senior managers: new research directions', *International Journal of Management*, vol. 21, no. 4, pp. 415–426.

Chesley, N. (2005) 'Blurring boundaries? Linking technology use, spillover, individual distress, and family satisfaction', *Journal of Marriage and Family*, vol. 67, no. 5, pp. 1237–1248.

Connors, E. (2006) 'BlackBerry in full flight in Australia', *Australian Financial Review*, 22 Feb., p. 52.

Cousins, K. C. & Robey, D. (2005) 'Human agency in a wireless world: patterns of technology use in nomadic computing environments', *Information and Organization*, vol. 15, no. 2, pp. 151–180.

Du Gay, P., Hall, S., Janes, L., Mackay, H. & Negus, K. (1997) *Doing Cultural Studies: the Story of the Sony Walkman*, Open University/Sage, London.

Duxbury, L. & Higgins, C. (2001) *Work–Life Balance in the New Millennium: Where are We? Where Do We Need to Go?*, Canadian Policy Research Network, Ottawa.

Duxbury, L., Higgins, C. & Lee, C. (1994) 'Work–family conflict: a comparison by gender, family type, and perceived control', *Journal of Family Issues*, vol. 15, no. 3, pp. 449–466.

Duxbury, L., Higgins, C. & Coghill, D. (2003) *Voices of Canadians: Seeking Work–Life Balance*, Human Resources Development Canada, Ottawa.

Economist Staff (2005) 'The CrackBerry backlash', *The Economist*, vol. 375, no. 8432, p. 78.

Fortunati, L. (2002) 'The mobile phone: towards new categories and social relations', *Information, Communication & Society*, vol. 5, no. 4, pp. 513–528.

Gilder, G. (1993) 'Metcalf's law and legacy', *Forbes ASAP*, 1 Sept.

Gordon, J. (2002) 'The mobile phone: an artefact of popular culture and a tool of the public sphere', *Convergence*, vol. 8, no. 3, pp. 15–26.

Govindaraju, M., Sward, D. (2005) 'Effects of wireless mobile technology on employee work behavior and productivity: an Intel case study', in *Designing Ubiquitous Information Environments: Socio-technical Issues and Challenges*, eds C. Sørensen *et al.*, Springer, New York, pp. 349–351.

Haddon, L. (ed.) (1997) *Communications on the Move: the Experience of Mobile Telephony in the 1990s*, CULCOM, Brighton.

Hallowell, E. (2005) 'Overloaded circuits: why smart people underperform', *Harvard Business Review*, vol. 83, no. 1, pp. 54–62.

Hatch, M. J. (1993) 'The dynamics of organizational culture', *Academy of Management Review*, vol. 18, no. 4, pp. 657–693.

Higgins, C. & Duxbury, L. (2005) 'Saying "no" in a culture of hours, money and non-support', *Ivey Business Journal Online*, July–August, pp. 1–5, Available at: http://www.iveybusinessjournal.com/view_article.asp?intArticle_ID = 566 (accessed 27 Mar. 2007).

Humphreys, L. (2005) 'Cellphones in public: social interactions in a wireless era', *New Media & Society*, vol. 7, no. 6, pp. 810–833.

Jarvenpaa, S. L., Lang, K. R., Tuunainen, V. K. (2005) 'Friend or foe? The ambivalent relationship between mobile technology and its users', in *Designing Ubiquitous Information Environments: Socio-technical Issues and Challenges*, eds C. Sørensen *et al.*, New York, Springer, pp. 29–42.

Katz, J. E. (ed.) (2003) *Machines that Become Us: The Social Context of Personal Communication Technology*, Transaction, New Brunswick, NJ.

Katz, J. E. & Aakhus, M. (eds) (2002) *Perpetual Contact: Mobile Communication, Private Talk, Public Performance*, Cambridge University Press, Cambridge.

Ling, R. S. (2004) *The Mobile Connection: the Cell Phone's Impact on Society*, Morgan Kaufmann, San Francisco.

Maiden, M. (2006) 'Midnight watchman', *Sydney Morning Herald*, 14 Apr., p. 31.

Mazmanian, M. A., Orlikowski, W. J. & Yates, J. (2005) 'Crackberries: the social implications of ubiquitous wireless e-mail devices', in *Designing Ubiquitous Information Environments: Socio-technical Issues and Challenges*, ed. C. Sørensen *et al.*, Springer, New York, pp. 337–344.

Middleton, C. A. & Cukier, W. (2006) 'Is mobile email functional or dysfunctional? Two perspectives on mobile email usage', *European Journal of Information Systems*, vol. 15, no. 3, pp. 252–260.

Middleton, C. A., Scheepers, H. & Cukier, W. (2005) 'Exploring the contradictions of mobility: a case study of BlackBerry users in Canada', in *Proceedings of the Hong Kong Mobility Roundtable*, eds J. Y. L. Thong & K. Y. Tam, HKUST, Hong Kong, pp. 299–309.

Morgan, G. (1997) *Images of Organization*, 2nd edn, Sage, Thousand Oaks, CA.

Nah, F., Siau, K. & Sheng, H. (2005) 'The value of mobile applications: a utility company study', *Communications of the ACM*, vol. 48, no. 2, pp. 85–90.

Nyíri, K. (ed.) (2003) *Mobile Democracy: Essays on Society, Self and Politics*, Passagen, Vienna.

Oksman, V. & Turtiainen, J. (2004) 'Mobile communication as a social stage: meanings of mobile communication in everyday life among teenagers in Finland', *New Media Society*, vol. 6, no. 3, pp. 319–339.

Parks, B. (2006) 'Addiction, withdrawal and the BlackBerry', *The Star-Ledger*, 29 Jan., p. 1.

Peters, O. & Allouch, S. B. (2005) 'Always connected: a longitudinal field study of mobile communication', *Telematics and Informatics*, vol. 22, no. 3, pp. 239–256.

Reguly, E. (2006) 'RIM on edge of big global growth story', *Globe and Mail*, 7 Mar., p. B2.

Rick Mercer Report. (2006) 'BlackBerry helmet', Available at: http://throwawayyourtv.com/2006/03/blackberry-helmet.html (accessed 27 Mar. 2007).

Schein, E. H. (2004) *Organizational Culture and Leadership*, 3rd edn, Jossey-Bass, San Francisco.

Senge, P. M. (1990) *The Fifth Discipline: the Art and Practice of the Learning Organization*, Currency Doubleday, New York.

Sharma, D. C. (2004) 'RIM takes BlackBerry to India', CNET News.com, Available at: http://news.com.com/RIM + takes + BlackBerry + to + India/2100-1041_3-5417375.html (accessed 27 Mar. 2007).

Sørensen, C. & Gibson, D. (2004) 'Ubiquitous visions and opaque realities: professionals talking about mobile technologies', *Info The Journal of Policy, Regulation and Strategy for Telecommunications*, vol. 6, no. 3, pp. 188–196.

Staw, B. M. (1981) 'The escalation of commitment to a course of action', *Academy of Management Review*, vol. 6, no. 4, pp. 577–587.

Tait, N. (2004) 'Judge backs blackberry manufacturer in dispute over patents', *Financial Times*, 3 Feb., p. 4.

Terry, J. A. (2006) '"CrackBerry" use increases despite patent dispute', *The State Journal*, 21 Apr., p. T6.

Van Den Berg, P. T. & Wilderom, C. P. M. (2004) 'Defining, measuring, and comparing organisational cultures', *Applied Psychology—An International Review—Psychologie Appliquée—Revue Internationale*, vol. 53, no. 4, pp. 570–582.

Waldie, P. (2006) 'RIM shares rise on upbeat results', *Globe and Mail*, 30 Jun, p. B1.

Feminizing the Mobile: Gender Scripting of Mobiles in North America

Leslie Regan Shade

A lively body of feminist research has examined how communication technologies have been gendered through their social uses—often unintended—and their design (Chabaud-Rychter, 1994; Cowan, 1985; Horowitz & Mohun, 1998; Lupton, 1993; Olzenziel, 1999). The telephone is particularly illustrative. Rakow (1992) and Moyal's (1992) research demonstrated that women's primary use of the telephone was for community bonding and family 'kin keeping' while Martin's (1991) study of Bell Telephone during the early twentieth century in Canada emphasized its move from a business tool for men to its feminization, achieved via women's labour as operators and their creation of a viable social culture.

Telephone technology and design have changed considerably to appeal to female consumers, reflecting its status as an indispensable domestic artefact, including stylistic trends (colours—plain black to pale hues), design (the Princess telephone) and technological innovations (push-button to portables) (Lupton, 1993). The mobile telephone is increasingly designed and marketed to appeal to women and female teenagers. Whether through the creation of features and accessories (ringtones, wallpaper, faceplates, camera phone, wireless synergies) or through branded phones of high-end fashion designers, the design of the mobile reflects a distinctive feminization. This mimics efforts in the1990s when media corporations and entrepreneurs feminized the Internet through the creation of popular content that privileged and encouraged women's consumption, rather than production or critical analysis (Shade, 2002).

Using the concept of the gender script (Rommes, 2002; van Oost, 2003), this paper sketches the feminization of the design and marketing of mobile phones in the North American context through an analysis of selected print advertisements appearing in women's and teen magazines. Despite their high penetration of broadband, Canada and the United States lag behind other developed countries in their uptake of mobile phones because both countries are well served by solid landline infrastructures. OECD statistics for 2004 indicate that the United States had 62 mobile subscribers per 100 inhabitants, and Canada 47; this contrasts with the OECD average of 72, with 109 in Sweden, 102 in the United Kingdom, 96 in Finland, and 72 in Japan (OECD, 2005). Unlike Europe and Asia, a distinctive mobile culture is nascent, and marketing the mobile for women and female teenagers in Canada and the United States is thus opportunistic and timely for mobile manufacturers and wireless carriers.

This paper first provides a brief overview of international scholarship on uses and perception of mobiles by women and teenagers. The next section explicates the gender script and examines selected print ads for mobile phones appearing in North American women's and teen magazines. How can we go beyond gender scripts that essentialize women and their uses of mobiles? The paper concludes with reflections towards this end.

Exploring Mobiles and Gender

Empirical studies on women and mobile phones focus on their everyday uses across cultural contexts. The ITU (2004) notes that mobiles are increasingly used by urban youth and women in developing countries, and are positioned for 'everyday life' through entertainment and lifestyle applications. Castells *et al.*, (2004) remark that, globally, mobile adoption amongst females is increasing, with subtle national differences in appropriation of services; Japan, South Korea, and China target female consumers through commercial campaigns and design of distinctive handsets and service plans (p. 152). Plant's research for Motorola illustrated how men and women use the mobile for public display—for competitive advantage, status symbol, or to signal self-containment; female teenagers 'are particularly aware of the fashion aspects of their mobiles, competing to acquire the latest, coolest models and to customize them in the latest, coolest ways' (Plant, 2003, p. 44). Customization includes ringtones, sticking logos and graphics on the phone, or attaching jewellery and trinkets.

Research on the creation of a distinctive youth mobile culture in Scandinavia and Japan (Caronia, 2005; Fortunati, 2002; Ling, 2004; Ito *et al.*, 2005; Oksman & Turtiainen, 2004) emphasizes a preoccupation with stylistic trends in design and applications, many coded as masculine and feminine. In Norway, Skog (2002) comments that 'the mobile phone companies seem to design phones to match the traditional female and the male cultures', while Wilska (2003) in Finland writes that 'trendy and impulsive consumption styles (that are usually regarded as "women's stuff") are connected to technology enthusiasm, which is usually seen as a "typical male" thing'.

To date, little scholarly research has looked at the marketing of mobiles to women or female teens in North America (unlike studies on Italy and Scandinavia; see Zatta, 2003 and Mörtberg, 2003). Townsend (2000), reviewing early US marketing, notes its importance in shaping ideologies of mobile usage; advertising to women was often marketed to dispel security and to promote domestic 'kin keeping', while for men it was 'most frequently marketed as a tool of fashion, power, and virility' (p. 6).

Gender Scripts

Van Oost describes a 'gender script' as:

> the representations an artifact's designers have or construct of gender relations and gender identities—representations that they then inscribe into the materiality of that artifact. Like gender itself, which is defined as a multi-level process, gender scripts function on an individual and a symbolic level, reflecting and constructing gender differences in the division of labor. (2004, p. 195)

Van Oost used this notion of gender scripts to describe the design history of shavers for men and women, highlighting how the artefacts embodied gender. Indeed, looking at current constructions of the razor we see how gender is overtly built into their design and marketing; just think of the steely-grey and green Mach3 turbo razor for men ('The best a man can get!'), versus the turquoise-blue Venus Divine shaver for women ('Reveal the Goddess in you!'). Perhaps coincidentally, Motorola introduced 'his' and 'her' slim Razr phones—his was jet black, while hers was a an 'eye catching tone of pink', and was soon, according to Motorola's promotional material

> seeded to a select group of fashion elite and Hollywood 'it' girls ... its shocking measurements make it one of the slimmest phones on the market. Rich functionality, performance excellence and design innovation give the user a total sensory experience—from photo messaging to hands free connections to superlative metallic finishes.
> (http://www.motorola.com/motoinfo/product/details/0,,128,00.html)

Rommes (2002) also theorizes gender script along with technological scripts and domestication as entry points to describe gendering through design, in her study of Amsterdam's community-based computer network The Digitale Stad (Digital City). Technological scripts are defined as 'the assumptions about the use context that are materialized in the technology, which pre-structure the use of technology ... they attribute and delegate specific competencies, actions, and responsibilities to their envisioned users' (Rommes, 2002, p. 15). When scripts reveal gendered patterns, they become gender scripts, which 'may emphasize or hide, and reinforce or diminish gender differences and gender inequalities' (Rommes, 2002, p. 18).

Domestication refers to how technology is incorporated into the everyday patterns of the users, and has been widely used by communication and feminist scholars to study the ways in which family members negotiate technologies within the household (Haddon, 2004; Silverstone & Hirsch, 1992). Adopting the four phases of domestication—appropriation, objectification, incorporation, and conversion—to mobiles provides a useful framework for exploring the gendering of mobiles. Appropriation occurs through the adoption of the mobile as a necessary domestic artefact; objectification through the personalization of the mobile; incorporation through its management of daily routines; and conversion as the mobile blurs the distinction between the private, domestic sphere and the wider public sphere. The mobile is thus situated instrumentally as a facilitator of social communication, particularly for women.

Gender scripts will be used to illustrate several trends with mobiles: the mobile as a new marketing terrain (especially targeting teenage girls) and the mobile as fashion accessory and branded haute couture. It should be emphasized that this paper focuses only on mobile phone design and marketing for women and female teenagers, and does not analyse how mobiles have been 'masculinized' as well via design and marketing.

Gendering and Marketing Mobiles

Mobile phones are considered the 'third screen' for marketing. Digital convergence includes text messaging, integrated content, games, interactive voice response, WAP sites, ringtones/ring-back tones, viral/geo-targeting/advertising on mobile broadcast, sponsorships/subsidizing cell phone costs, and mobile telemarketing (May & Hearn, 2005). Gartner Research has estimated that Americans spent $1.2 billion in 2004 on ringtones, wallpaper and personalization services and $1.4 billion on cell phone games and entertainment (Thottam, 2005). Given their purchasing power, teenagers are the target for many of these features, with mobile media poised to 'turn consumption into both a fashion statement and an impulse purchase' (Siklos, 2006, p. B3). Advertising imagery and copy positions the converged mobile as a unique signifier for expressing one's individuality through customization of features and design. Personalization is present in many advertisements, and reflects both appropriation and objectification through advertising scripts and imagery.

In Canada, Rogers Wireless teamed with MuchMusic (the Canadian equivalent of MTV), for the MuchMusic Edition Phone, with content including 'exclusive' private access to MuchMusic updates, a videophone to send and receive video clips, a TuneTracker ('hold your phone up to the music for the song's name and artist') and the MuchMusic Factory, to create individualized ringtones. 'Finally, a phone as individual as you are', reads the ad copy, depicting a young teenage girl dancing solo on a stage.

Sprint's marketing of ringtones via their BlingTones trademark features new recordings by hiphop producers. An ad displays four exhausted women shoppers resting on a couch, shot from their necks down, laden with colourful shopping bags. Ad copy reads: 'You've got the look. Now what about the sound? Make your phone your own. It's easy.'

Nokia is a pioneer in fusing functionality, usability and fashion. Their 7280 'Lipstick Phone' was heralded for being 'small, sexy, eye-catching and cool' (Thompson, 2005, p. 233). Nokia welcomes customization of their phones with covers, wallpapers, and 'rhythmic ringing lights'. The Nokia 7270 folding phone features 'chic, interchangeable fabric wraps... allow(s) you to impulsively change your look as often and as boldly as you please. Stay ahead of the latest trends, or simply set them yourself.' Or, one can customize a phone with faceplates: 'How will you express yourself', the ad copy reads, with a photo of a young woman staring out at us, resting on her bed, with a matching shirt and faceplate. Another ad contrasts a silver phone with a nearly identical silver shoe: 'Accessorize accordingly' reads the copy.

Samsung's videophone downloads, plays and e-mails video. In one ad a young woman displays the phone against a neon movie theatre marquee that reads 'Cinema Village'. Says the ad copy: 'Your life. Now showing ... allows one to record 15 seconds of digital video and audio as you walk along the red carpet of your life.'

LG, a relatively new entrant to the mobile market, promoted its camcorder phone through ads that situate the phone as a fashion accoutrement. One ad shows a woman in a public washroom looking disdainfully away from another woman reflected in the mirror above the sink, remarking: 'I can't believe she has on the same phone as me.' Another ad shows a young man at a club, with an unremarkable pick-up line to a blonde woman, presumably checking messages on her mobile: 'That phone really brings out your eyes.'

Incorporation and conversion are reflected in ads that highlight the mobile as a tool to coordinate social engagements. mMode from AT&T Wireless combines wireless Internet with e-mail, instant messaging, games, and various content (news, weather, sports, entertainment, lifestyle, games, pictures, travel and dining). An ad displays a woman seated at a bar stool looking at her mobile, with a smiling man behind her. Says the ad copy: 'You can use mMode to make a connection on Match Mobile, then choose a restaurant and find directions to get you there. It will even help you find the nearest ATM in case you want to go Dutch.' Samsung's A800 phone is the 'perfect dining and traveling companion—it's sleek, stylish, powerful, and it never forgets a face or a name'. Besides being a camera phone, it takes dictation 'and you don't even have to buy it a drink'.

Incorporation and conversion are also in play via mobile-casting; one particular example is a venture between Mott's (an apple and fruit juice company owned by parent Cadbury Schweppes) and Epicurious.com (the recipe website for *Gourmet Magazine*) wherein 'Mott's recipe ideas and made-to-order shopping lists [are] beamed directly to your mobile phone.' The ad in *Bon Appétit Magazine* shows a white, blonde woman in breezy summer white attire, shopping bags in hand, smiling blissfully at her mobile.

Marketing the Mobile to Tweens

Teenagers are considered the 'sweet spot' for mobile designers. US statistics estimate that '76% of young people ages 15 to 19, and 90% of people in their early 20s regularly use their cell phones for text messaging, ringtones, and games, and that enthusiasm has turned wireless data services into a significant business' (Thottam, 2005).

Celebrity branding of mobiles for female tweens is one such trend. Single Touch pioneered the Hilary Duff phone, a prepaid handset featuring preloaded wallpaper and ringtones exclusive to the teen star. Boasting itself as the first phone catering to pre-teen girls, Mattel and Nokia launched the Barbie phone packaged with prepaid minutes, customized ring tones and faceplates (Heller, 2005). Marketed alongside the Barbie MyScene toy line, the phones are sold in US mega-retail stores Target, Toys R Us, and WalMart. Timothy J. Kilpin, senior vice-president for girls' marketing and design at Mattel, claims the Barbie phone 'transcends traditional doll play—it's steeped in entertainment. It allows girls to interact with the brand' (Crockett & Kharif, 2005). The Nokia 3587i phone features a full-colour screen and voice dialling. Retailing for US$49.95, it comes with 30 minutes of airtime, with additional minutes sold in $5 increments. Mattel and Single Touch Interactive are creating content for the phone, including nine preloaded MyScene wallpapers, three 'rockin' ringtones', and three faceplate designs. Promotional material positions the phone as a parenting device, deviously aligning its use to rewarding proper behaviour from the child, while maintaining a panoptic gaze:

> Parents not only have the peace of mind knowing that they are 'Always in touch' with their child but they can also stay in control of phone usage by tying their child's good behavior to added minutes via the Reward Board. Just go online [to Single Touch's http://www.myscenemobile.com/], set a list of chores which can include making the bed, finishing homework or not arguing with your brother or sister, and place stars on the completed tasks. At the end of the week or month, parents can buy extra minutes according to the child's list of completed tasks. (http://www.singletouch.net/brands/msm.htm)

Similarly, MGA's popular Bratz line of nubile dolls has coordinated with Sony Ericsson for the 'Bratz Mobile'; features allow parents to control how their children use the phone through restricting outgoing calls to pre-programmed and approved numbers only.

The Mobile as Fashion Accessory and Branded Haute Couture

Lacohée *et al.* (2003) argue that 'the idea that the mobile is on constant show and is therefore a fashion accessory has fed into an advertising rhetoric of continual upgrading to avoid being shamed' (p. 208). This sentiment was echoed by rapper mogul Sean 'P. Diddy' Combs in his comments to the Cellular Telecommunications & Internet Association convention, where he mused on the blending of hip-hop and cell phone culture: 'It's not about the minutes or the service ... it's about how the person is defined' (Rhoads, 2005). Likewise, Munich-based fashion house Escada spokeswoman Sabine Eisenreich enthused that the mobile 'is not just a functional thing ... it's become an accessory, like shoes, bags, belts or jewelry' (Rhoads, 2005).

Industry market research is aligned with Eisenreich's sentiments; new points of sale for mobile handsets include upscale fashion boutiques that merge haute couture designers with handset companies: Siemens/ESCADA, Nokia/Versace, Nokia/Vertui. Motorola and Samsung have coordinated with designers Diane Von Furstenberg, Anna Sui and Vivienne Westwood. Strategies utilized in the haute couturing of the mobile include co-branding of handset vendors with designers, and handsets incorporating fashion elements. Siemens and Escada launched three limited-edition cell phones; one version, Denim and Diamonds, was decorated in denim, Swarovski crystals and a string of attached pearls. According to an Escada spokeswoman, all 7,000 of the US$1,040 handsets sold out within two months (Rhoads, 2005).

Samsung is a vanguard designer for feminized mobiles. In 2002 it launched the red 'Ladyphone'—the design mimicked a make-up compact, featuring 'a biorhythm calculator, a fatness function that calculates a user's height-to-weight ratio, a calendar for keeping track of your menstrual cycle and a calorie-counting function. Enter an activity (cleaning, dishwashing, cooking, shopping) and the time spent, and the phone works out how many calories have been consumed' (The Economist, 2002). Samsung promotional copy described the Egèo phone as 'engraved with a gold rose ... a beautiful symbol of your aristocratic dignity. Its small and compact design adds a warm glow in your hands like a precious gem' (see http://www.samsung.com/my/products/gsm/gsm/sgh_a400.asp). Samsung also sponsored the 2002 Britney Spears concert tour. Designer Diane von Furstenberg, exclaiming that the mobile phone had become 'part of a woman's body language' (Rhoads, 2005), designed her Samsung namesake mobile with a replica of her 1974 Andy Warhol portrait.

In 2004 the *Wall Street Journal* reported that the Korean company was 'the world's No. 1 purveyor of high-end cellphones and the third-largest seller overall' (Choi, 2004). Lee Ki Tae, head of Samsung's handset business, emphasizes the everyday ubiquity of the mobile: 'There isn't a single other electronic gadget you have around 24 hours a day, every day of the year ... Consumers are developing a strong emotional bond with their handsets' (ibid.). Reportedly more than 100 designers work for Samsung, and their innovative marketing strategy has showcased their products in the Guggenheim Museum and Versailles Palace.

The limited edition (1,000 sets were manufactured) Anna Sui Mobile, a combination camera/camcorder phone, was launched in conjunction with *Vogue Magazine*. Said Sui: 'The wireless phone is the most visible accessory and should be an extension of your personal style and essence.' A fashion spread shows a model with a Martini examining her mobile; the copy reads: 'He's running late, so e-mail him a picture with your Anna Sui Mobile by Samsung and show him what he's missing.'

Designer Betsey Johnson, known for her flamboyant and bohemian clothes, also teamed with Samsung to create a handset and accessories ('comes complete with a matching bag for you to flaunt around town'). Said Johnson: 'It's all about personality—you need to look for a phone with a personality that you enjoy—that matches you.'

Conclusion: Mobilizing Gender?

Gender scripts can 'illuminate how gendered user representations are an inextricably part of designing artifacts' (Van Oost, 2003, p. 194). The description of mobile advertising in this paper highlights how women have been inscribed as particular users and consumers. Mobiles were first marketed in the 1990s to women as a tool for 'remote mothering' (Rakow & Navarro, 1993) or as security and safety devices for travel. As mobile penetration increases in North America—particularly amongst young women and tweens, entrepreneurs and telecom companies are counting on cross-media convergence to spark new marketing platforms on the 'third screen', while they are also promoted as branded fashion accessories and as status symbols through limited edition haute couture items. The gendered scripts here, through explicit user-representations, reinforce women's role as consumer and as object, with uses for sociability, shopping and entertainment. As Kearney argues in her analysis of popular film and television programmes in post-Second World War America, the common trope of the female teenager on the telephone was 'not part of an attempt to *eliminate* teenage girls' independence altogether, as one focused on preventing its further subversion of heterosexual patriarchy' (Kearney, 2005, p. 594). Likewise, contemporary marketing of the mobile in North America reinforces femininity and hetero-normativity; the mobile is an extension of one's stylistic sensibility, the gaze is narcissistic and competitive. These instrumental uses certainly contrast with the innovative keitai culture in Japan (Ito *et al.*, 2005).

Jain reminds us to pay attention to 'the ways in which objects such as cell phones come with their own disciplining networks' (Jain, 2002, p. 424). Similar to technological scripts, disciplining is seen starkly in many of the mobiles for tweens, from parental controls that limit outbound calling to restricting the length of chat via pre-paid calling cards. Samsung's Egèo phone, with its 'Pink Schedule' (menstrual calendar), Bio Rhythm, and Calories and Fatness Counter also discipline, monitor, and essentialize women through emphasizing bodily preoccupations. Camera and videophones, text messaging, ringtones, and e-mail are features built into mobiles, marketed for women in their roles as consumers and girlfriends.

Jain also challenges to us to think critically about the engendering of mobiles, 'and what these in some ways radically new embodiments mean in our thinking on subjectivity, civic action and community' (Jain, 2002, p. 424). Are there ways that mobiles have been used for cultural production or civic participation, and are there inventive designs that could facilitate such usage?

The case of the Village Phone project in rural Bangladesh funded by the Grameen Bank has become emblematic of ICTD success stories; Aminuzzama *et al.*'s (2003) assessment highlights how married 'phone ladies' are mostly the owners, whereas most of the users are men. And, although mobiles are promoted for 'networked individualism' (Miyata *et al.*, 2005), Weilenmann & Larsson (2000) have observed that teens use their mobiles collaboratively and for social purposes, and argue that 'it is important to ground design of new mobile technology for young people, in an understanding of how teenagers' actually use their mobile phones in their everyday lives' (p. 2). Lee (2005) describes how young South Korean women appropriate the camera phone for cultural production, despite the prevalence of advertising that shows men snapping pictures of women: 'these women are not the mere owners of camera phones, but performers who create various cultural meanings. They develop a more intimate relationship with technology, challenge the conventions of gaze, give meaning to what is taken, and circulate their own expressions.'

While not disavowing that fashion trends are an intrinsic facet of identity formation (particularly for young women), the challenge now is how to design mobile phones (which are undoubtedly used by many women as a mundane tool both to support domesticity and supplement the increasingly mobile workplace) to more seamlessly integrate domestic functionalities alongside creative, cultural, playful, and perhaps even civic modes of communicating. Bell *et al.*, (2005) argue that defamiliarizing the known within the domestic context can provoke critical and culturally sensitive perspectives on technology design. Becoming cognizant of how design and marketing incorporate stereotyped gender assumptions through gender scripts is important. While the gender scripts analysed in this paper accentuate emancipatory uses, other socio-technical networks may conceive of the mobile as less liberating. Where and how are women involved in mobile phone design, production, and manufacturing? Feminist interrogations of mobiles must thus also go beyond mere critical assessments of the consumption of mobiles, and focus on the political economic realities in their production.

References

Aminuzzama, S., Baldersheim, H. & Jamil, I. (2003) 'Talking back! Empowerment and mobile phones in rural Bangladesh: a study of the Village Phone scheme of Grameen Bank', *Contemporary South Asia*, vol. 12, no. 3, pp. 327–348.

'ARC chart report reveals fashion label handsets increase in popularity'. (2005) *Telecomworldwire*, 13 Jun.

Bell, G., Blythe, M. & Sengers, P. (2005) 'Making strange: defamiliarization and the design of domestic technologies', *ACM Transactions on Computer–Human Interaction*, vol. 12, no. 2, pp. 149–173.

Caronia, L. (2005) 'Mobile culture: an ethnography of cellular phone uses in teenagers' everyday life', *Convergence*, vol. 11, no. 3, pp. 97–103.

Castells, M., Fernandez-Ardevol, M., Linchuan Qui, J. & Sey, A. (2004) *The Mobile Communication Society: a Cross-cultural Analysis of Available Evidence on Uses of Wireless Communication Technology*, USC Annenberg Research Network on International Communication, Available at: arnic.info/workshop04/MCS.pdf (last accessed 1 Sep. 2006).

Chabaud-Rychter, D. (1994) 'Women users in the design process of a food robot: innovation in a French domestic appliance company', in *Bringing Technology Home: Gender and Technology in a Changing Europe*, eds C. Cockburn & R. Furst Dilic, Open University Press, Buckingham, pp. 77–93.

Choi, H. W. (2004) "How do you make a "premium" cellphone?", *Wall Street Journal*, vol. 15, p. B1.

Cowan, R. S. (1985) *More Work for Mother: the Ironies of Household Technology from the Open Hearth to the Microwave*, Basic Books, New York.

Crockett, R. O. & Kharif, O. (2005) 'Calling preteens with a Barbie phone', *Business Week* [online], 18 Feb., Available at: http://www.businessweek.com/technology/content/feb2005/tc20050218_4609_tc024.htm

The Economist. (2002) 'From cell phones to self phones', 24 Jan., Available at: http://www.economist.com/business/displayStory.cfm?Story_ID=954329

Fortunati, L. (2002) 'The mobile phone: towards new categories and social relations', *Information, Communication & Society*, vol. 5, no. 4, pp. 514–528.

Haddon, L. (2004) *Information and Communication Technologies in Everyday Life: a Concise Introduction and Research Guide*, Berg, London.

Heller, L. (2005) 'Back-to-school: pencils, homework, cell phones?', *DSN Retailing Today*, vol. 44, no. 8, p. 42.

Horowitz, R. & Mohun, A. (eds) (1998) *His and Hers: Gender, Consumption, and Technology*, University Press of Virginia, Charlottesville.

International Telecommunication Union (ITU) (2004) *Social and Human Considerations for a More Mobile World: Background Paper*, 24 Feb., Document SMIS/04, ITU/MIC Workshop on Shaping the Future Mobile Information Society, Seoul, 4–5 March.

Ito, M., Okabe, D. & Matsuda, M. (eds) (2005) *Personal, Portable, Pedestrian: Mobile Phones in Japanese Life*, MIT Press, Cambridge, MA.

Jain, S. S. L. (2002) 'Urban errands: the means of mobility', *Journal of Consumer Culture*, vol. 2, no. 3, pp. 410–438.

Kearney, M. C. (2005) 'Birds on the wire: troping teenage girlhood through telephony in mid-twentieth century US media culture', *Cultural Studies*, vol. 19, no. 5, pp. 568–601.

Lacohée, H., Wakeford, N. & Pearson, I. (2003) 'A social history of the mobile telephone with a view of its future', *BT Technology Journal*, vol. 21, no. 3, pp. 203–211.

Lee, D. (2005) 'Women's creation of camera phone culture', *Fibreculture Journal*, no. 6 Available at: http://journal.fibreculture.org/issue6/issue6_donghoo.html.

Ling, R. (2004) *The Mobile Connection: the Cell Phone's Impact on Society*, Morgan Kaufmann, New York.

Lupton, E. (1993) *Mechanical Brides: Women and Machines from Home to Office*, Cooper-Hewitt National Museum of Design, Smithsonian Institute, and Princeton Architectural Press, New York.

Martin, M. (1991) *Hello Central? Gender, Culture, and Technology in the Formation of Telephone Systems*, McGill-Queen's University Press, Montreal.

May, H. & Hearn, G. (2005) 'The mobile phone as media', *International Journal of Cultural Studies*, vol. 8, no. 2, pp. 195–211.

Miyata, K., Boase, J., Wellman, B. & Ikeda, K. (2005) 'The mobile-izing Japanese', in *Personal, Portable, Pedestrian: Mobile Phones in Japanese Life*, eds M. Ito, D. Okabe & M. Matsuda, MIT Press, Cambridge, MA, pp. 143–164.

Mörtberg, C. (2003) 'Heterogeneous images of (mobile) technologies and services: a feminist contribution', *NORA*, vol. 3, no. 11, pp. 158–169.

Moyal, A. (1992) 'The gendered use of the telephone: an Australian case study', *Media, Culture and Society*, vol. 14, pp. 51–72.

Oksman, V. & Turtiainen, J. (2004) 'Mobile communication as a social stage: meanings of mobile communication in everyday life among teenagers in Finland', *New Media & Society*, vol. 6, no. 3, pp. 319–339.

Olzenziel, R. (1999) *Making Technology Masculine: Men, Women and Modern Machines in America, 1870–1945*, Amsterdam University Press, Amsterdam.

Organization for Economic Cooperation and Development (OECD) (2005) *OECD ICT Key Indicators*, Available at: http://www.oecd.org/sti/ICTindicators

Plant, S. (2003) *On the Mobile: the Effects of Mobile Telephones on Social and Individual Life*, Motorola Corporation.

Rakow, L. (1992) *Gender on the Line: Women, the Telephone, and Community Life*, University of Illinois Press, Chicago.

Rakow, L. & Navarro, V. (1993) 'Remote mothering and the parallel shift: women meet the cellular telephone', in *Critical Studies in Mass Communication*, 10 Winter, pp. 144–157.

Rhoads, C. (2005) 'Cellphone makers, users heed fashion world's call', *Wall Street Journal*, 18 Mar. p. B1.

Rommes, E. (2002) *Gender Scripts and the Internet: the Design and Use of Amsterdam's Digital City*, Twente University Press, Enschede.

'Samsung launches Anna Sui couture phone' (2005) *Wireless News*, 21 Feb., p. 1
Shade, L. R. (2002) *Gender and the Social Construction of the Internet*, Peter Lang, New York.
Siklos, R. (2006) 'How much profit is lurking in that cellphone', *New York Times*, 5 May, pp. B3.
Silverstone, R. & Hirsch, E. (eds) (1992) *Consuming Technologies: Media and Information in Domestic Spaces*, Routledge, New York.
Skog, B. (2002) 'Mobiles and the Norwegian teen: identity, gender, and class', in *Perpetual Contact: Mobile Communications, Private Talk, Public Performance*, eds J. E. Katz & M. Aakhus, MIT Press, Cambridge, MA, pp. 255–273.
Thompson, H. (2005) *Phone Book*, Thames & Hudson, London.
Thottam, J. (2005) 'How kids set the (ring) tone', *Time*, vol. 165, no. 14, p. 40.
Townsend, A. M. (2000) 'Life in the real-time city: mobile telephones and urban metabolism', *Journal of Urban Technology*, vol. 7, no. 2, pp. 85–104.
Van Oost, E. (2003) 'Materialized gender: how shavers configure the users' femininity and masculinity', in *How Users Matter: The Co-construction of Users and Technology*, eds N. Oudshoorn & T. Pinch, MIT Press, Cambridge, MA, pp. 193–208.
Weilenmann, A. & Larsson, C. (2000) 'Collaborative use of mobile telephones: a field study of Swedish teenagers', in *NordiCHI2000 Proceedings*, Stockholm, 23–25 October 2000, Available at: www.viktoria.se/~alexandra/publications/collaborative%20use.PDF (last accessed 21 Mar. 2007).
Wilska, T. (2003) 'Mobile phone use as part of young people's consumption styles', *Journal of Consumer Policy*, vol. 26, no. 4, pp. 441–463.
'Worldwide mobile phone sales up 30% in Q4' (2005) *Wireless Week*, 15 Mar. vol. 11, no. 7, p. 64.
Zatta, S. (2003) 'Powerful objects for powerful women? Feminine representation in Italian technology advertising imagery', Paper presented at 'Gender and Power in the New Europe', *5th European Feminist Research Conference*, Lund University, Sweden, p. 20–24 Aug.

'What Hath God Wrought?' Considering How Religious Communities Culture (*or Kosher*) the Cell Phone

Heidi Campbell

Since Samuel Morse posed his famous question 'what hath God wrought?' in the first telegraph conversation, communication technology has been infused with spiritual undertones. The idea of a disembodied voice speaking across time and space echoes an otherworldly quality that remains in many reflections on this technology (see 'A spiritual connection', 2005). Similarly, in the past five years numerous religious groups have embraced and adapted features of the cell phone for religious purposes. From the Pope's daily religious text and prayer messaging services ('Pope spreads the word by SMS', 2003) to Hindus in Bombay avoiding quest at temples by sending prayers to Ganesh via SMS ('Hindus make their offerings by SMS', 2003). For a small fee British Muslims can receive verses from the Koran and fulfil their religious duties to give alms to the poor through a 'TXT & Donate Islamic Prayer alert' (Biddlecombe, 2004). Religious groups have also become innovators, designing special phone features to enable religious devotion. Muslims can purchase handsets with a built in 'Mecca Indicator' that also reminds them of prayer times (Biddlecombe, 2004). A Taiwanese handset maker designed a phone dedicated to the Chinese goddess Matsu, which features a hologram of the goddess, religious chanting ring tones and is reportedly blessed in a special ritual at a Matsu temple ('Phones have been blessed . . .', 2004). Yet the mixing of religion with the cell phone has also been controversial in some cases. SMS greetings replacing traditional visits from friends and family during Ramadan have received public criticism in the Middle East (Qusti, 2003) and Catholic Bishops in the Philippines have publicly denounced wireless confessional services accepting confession or offering absolution via SMS ('Virging mobile users . . .', 2004).

Yet use may become problematic when the technology encroaches on traditional religious boundaries or is seen as an entity promoting an unwanted cultural or values shift. In light of this, religious communities often undergo a complex communal reflection and negotiation related to that technology. This communal negotiation raises interesting questions: how do religious groups evaluate new forms of technology, such as the cell phone? What processes come into play in their decision making about whether to accept, reject or alter the cell phone?

This article will look at how religious communities have responded to the cell phone by exploring the emergence of the kosher cell phone in Israel. In order to do this we must first discuss the process that religious user communities undergo when negotiating with technology. This is done through a social shaping of technology (SST) approach illustrated in Zimmerman-Umble's study of the Amish and the phone (Zimmerman-Umble, 1992). This provides a framework for evaluating cell phone use, highlighting the roles that resistance, reconstruction and rhetoric play in religious response to a technology. Finally this is applied to the case of the kosher phone.

Religious Negotiation with Communication Technology: Amish and the Phone

While the specific study of how religious communities negotiate with communication technology is an underdeveloped area, it has much resonance with the SST approach. This approach sees technological change and user innovation as a social process. It studies how and why a community of users responds to a technology in a certain way and helps to identify what values or beliefs influence this negotiation. One application of SST is the 'domestication' of technology, presented by Silverstone *et al.* (1992). They argue that technologies are conditioned and tamed by users in ways that enable them to fit more neatly into the routines of daily life or 'the moral economy of the household'. Moral economies are distinct spaces where symbolic meaning transactions occur. Domesticating a technology means making choices about the meaning and practice of a technology within this sphere. Thus a technology is shaped by the setting in which it lives and by the agents who utilize it. It is being argued that religious user communities can be seen as a 'family of users' who create a distinctive 'moral economy' of social and religious meaning which guides their choices about technology and rules of interaction with them. Uncovering this process can provide valuable insights into understanding how religious groups negotiate their technological use.

One of the few SST studies of religious negotiation with a technology is noteworthy for this investigation—the case of the Amish response to the telephone. Zimmerman-Umble's study details how the telephone was banned from Amish homes in Pennsylvania in 1909, solidifying most outsiders' view of the Amish as a group who rejects all modern technologies. However, she points out that while the telephone does not appear in the home, the Amish have not completely rejected its use. Instead they have created 'community telephones' located in 'shanties', often located at the intersection of several Amish farms. Thus it is not a blanket rejection of technology but a decision to require phones to be used and accessed communally.

The Amish response to the telephone is framed by their beliefs about communication and social life. Their use of technology is 'structured through the rituals of community life and anchored in the home' (Zimmerman-Umble, 1992, p. 185). The home is the centre of community life, where most social interaction occurs, including Sunday services, weddings and funerals. The introduction of the telephone into the home in the early 1900s presented a challenge to Amish core social patterns and values. It provided important access to current information and emergency services, but also facilitated association with outsiders, promoted individualism and drew members into private gossip, thereby encouraging disharmony.

After church elders discussed the benefits and problems that the telephone created it was decided that the telephone was not a 'necessity' '... [it] did not conform to time-honored principles of nonconformity and separation from the world' (Zimmerman-Umble, 1992, p. 189). This was a controversial decision, which some sources feel contributed to a later split in the community (Zimmerman-Umble, 1992, pp. 189–190). However, this was not a complete prohibition; it only prohibited private ownership. Home phones were seen to 'spoil the natural rhythm of family life', but community phones kept outside provided access in ways that monitored use and stressed face-to-face contact as preferred communication. Over time the Amish community has modified its standards, including allowing telephones in small businesses or workshops, yet the desire to maintain community still influences whether such changes should occur.

Zimmerman-Umble highlights two important lessons from the Amish response to the telephone. First, a social group's historic and cultural orientation shapes the meaning they give to a technology and how they respond to it. She argues that the telephone as a technology 'has little universal meaning apart from that which is constructed or negotiated by those social groups who make use of it ... meaning is transformed as social and cultural boundaries are crossed' (Zimmerman-Umble, 1992, p. 183). This highlights the idea of resistance to technology, which Zimmerman-Umble mentions in her chapter title. Technologies come with benefits and costs, and embracing a technology may mean accepting certain standards of use. Resistance involves a social group choosing not to reject a technology but instead to resist certain aspects or outcomes. Resistance means a community becomes actively involved in a process of analysing the technology or negotiating standards of acceptable use.

Second, she states that studying groups who reject or reconfigure a communication technology provides important insights into assumptions about the social meaning of technologies. 'The Amish rejection of the communication technologies such as the telephone seems to articulate distinct social boundaries and in turn facilitates maintenance of those boundaries and the community as a whole' (Zimmerman-Umble, 1992, p. 183). Here values or social objectives dictate a technologies role and relation to the community. If the technology is seen as valuable, but incompatible in its current form, reconstruction must take place. Reconstruction involves making choices about what form of the technology will be acceptable, the context in which it should be used, how it should be utilized and if any aspect must be modified. So if a social group decides that it must resist certain aspects of a technology in order to maintain its social boundaries the technology must be reconstructed so it can be integrated into community life.

These processes of resistance and reconstruction can be further seen in Amish negotiation with the cell phone. Rheingold's investigation of Amish cell phone use points out that:

> New things are not outright forbidden, nor is there a rush to judgement. Rather technologies filter in when one of the more daring members of the community starts use, or even purchase something new. The others try it. Reports circulate about the results. What happens with daily use? Does it bring people together? Or have the opposite effect. (Rheingold, 1999)

The cell phone again highlighted these questions for community members and leaders, while raising new concerns. As a technology created to be personal and portable it prompted fears that the cell phone would encourage assimilation of values outside the community, including individuality and personal control over community responsibility. Rheingold, quoting Zimmerman-Umble, states: 'what makes the cell phone so handy—the lack of a wire—also poses a special challenge for the Amish'. The Amish rely only on electricity they create themselves, run by diesel generators or powered by hydraulics, so they do no have to connect with public utilities outside the community. Wires mean the community can easily see who is using the electricity source or is connected to the telephone, but the wireless cell phone removes this ability to monitor use. According to Rheingold, some Old Order Amish leave cell phones in community shanties to address this issue, while others leave them overnight with an 'English' neighbour who recharges it for them for pick-up in the morning. He does document a few cases of personal ownership, but highlights the dilemma this raises as it seems to promote the rights of the individual in a social context where community comes first.

Framework for Evaluating Religious Cell Phone Use

A social shaping approach to religious user communities is being argued as a valuable way to study religious engagement with technology. It is shown above that Zimmerman-Umble's utilization of domestification highlights how religious communities may tame a technology so that it can fit more acceptably into the private, tightly bounded social sphere of the community. Similarly, Barzilai-Nahon & Barzilai (2005) suggest that in order to understand how religious groups interact with new technologies one must study how they modify or 'culture' a technology in ways that preserve, rather than subvert, their unique culture. Through examining the ultra-Orthodox community in Israel, a highly conservative sect of Judaism, and its use of the Internet they found a complex social value-construction and negotiation process taking place between religious users and leaders. While ultra-Orthodox rabbis initially condemned Internet use, because it allowed unmonitored secular content into the community, many members wanted to use the Internet for economic purpose and self-expression. In order to make their case, users, especially females, described the Internet in terms of needs, emphasizing how the technology met certain needs within the community. Women are not only the primary care givers in the ultra-Orthodox community but also work to support the family so men can fulfil their requirement of daily prayer and Torah study. The Internet became valued for enabling these women to work at home. This recognition led to changes in official views within the community about the technology. Barzilai-Nahon and Barzilai identify the end result of this negotiation process as a 'cultured technology'. For them, culturing a technology involves making the technology acceptable within a communal context by reshaping it through use and discourse so it is in line with the valued ideals of the group. The extent to which a religious group can culture—frame, modify or re-design—a technology determines to what extent it can be deemed 'culturally appropriate' (Barzilai-Nahon & Barzilai, 2005, p. 25) and thus incorporated into the life of the community.

Barzilai-Nahon and Barzilai's concept of 'cultured technology' in some respects simply echoes Zimmerman-Umble's interpretation of domestification, as both approaches emphasize that religious groups with strict social boundaries and patterns undergo a distinctive negotiation process with technologies in light of community values and practices. Zimmerman-Umble's use of domestification highlights the need to identify the meaning bestowed on a technology by a group's distinct culture and history; this determines the social controls and value judgements it exerts on that technology. She urges us to identify the moral economy of the community and how this may call for the reconstruction of a technology that is deemed valuable, so its use or design come into agreement with community life and practices. Barzilai-Nahon and Barzilai's culturing approach places emphasis on how a community's discourse frames their approach to a technology and how they may employ different forms of rhetoric or argumentation to negotiate and eventually 'culture' a specific technology. This means highlighting community rhetoric surrounding the decision-making process in relation to the technology. Here we will take a combined approach by discussing how a community's history and culture guides practices of resistance, reconstruction and their use of rhetoric to frame the technology. This framework is now applied to a unique example of community negotiation—the rise and culturing of the 'kosher' cell phone.

Case Study: the 'Kosher' Cell Phone

In March 2005 MIRS Communication, an Israeli wireless company, announced the launch of a phone designed specifically for the ultra-Orthodox Jewish community in Israel. The idea for such a phone began when religious authorities and community members became concerned that cell phones might be enabling unacceptable content to infiltrate the community. This led to an official, lengthy and interesting process of value and need negotiation between the community and several mobile phone providers in 2004 and 2005. The result was the launch of the 'kosher phone', a unique example of how a religious community may re-design a technology so that it is compatible with its standards of practice.

The initial 'kosher' (meaning approved or acceptable under rabbinical, religious law) phones were first-generation Motorola handsets that had been modified to disable Internet access, SMS text messaging and video and voice mail application. These are services that many rabbinical authorities felt would expose members of this conservative religious community to dubious, unmonitored secular content. These phones are visibly marked with a stamp signifying approval by rabbinical authorities. *Mobiledia*'s online reports described MIRS efforts as a decision 'to cater toward a demographic largely neglected by major cell phone companies; the ... ultra-Orthodox Jews' where these features are 'deemed threatening to their conservative way of life, particularly with the young' ('Kosher phones for ultra-Orthodox Jews', 2005).

Ultra-Orthodox Judaism is a small conservative segment of the overall religious community of Israel, representing about 6 per cent of the total population. Also referred to as haredi (or charedi), the community is highly religious and characterized by its rejection of the values of modernity, following kosher practices that seek to separate the sacred from the profane in all areas of life, a commitment to religious study and prayer, and the wearing of the dress and head coverings of their ancestors of eighteenth-century Europe (Heilman & Friedman, 1991). It is important to note that ultra-Orthodoxy differs from other forms of Judaism (Stadler, 2002), such as the more liberal expressions of conservative and reformed Judaism dominant in the United States, or Orthodox Judaism, which is the most prevalent expression in Israel. The ultra-Orthodox have been described as a 'culture of the enclave', a community which follows strict religious rules along with an isolated lifestyle (Stadler, 2005, p. 217). Most Orthodox Jewish communities typically take a concerned and reflective approach towards media technology. Yet the ultra-Orthodox often characterize media technologies (i.e. TV and the Internet) as symbols of modernity and secular values, traits from which they consciously distance themselves. Thus use of such technology is often debated within the community and the subject of law making. Discussions about technology use facilitated by rabbis and religious leaders are often framed in terms of the possibilities and dangers that the technology affords the community. Public discussions related to the boundaries of use of certain modern technologies are common. For example, how electrical appliances might be used on Shabbat, the Jewish sacred day of the week (lasting from sunset to sunset, Friday to Saturday) to reasons for forbidding televisions in homes, as it symbolizes the epitome of secular values, seen as a device aimed at moral seduction (Stadler, 2005).

Ultra-Orthodox religious leaders in Israel lauded the launch of the kosher phone as a success and a demonstration of the 'consumer power of the charedi community' ('Opinion & comments: kosher cell phones', 2005). 'Rabbonim and public figures said *maran verabonon* (masters and sages) defined the battle for kosher communication as an existential battle, a battle for the soul, and that every effort must be made to insure its success' (Kahn, 2005).

The kosher phone represents a unique negotiation between a religious user community and a cellular phone company ('Company markets kosher cell phone', 2006), resulting in a technology cultured towards specific religious ideals. However, this is not simply the story of a service provider meeting consumer demands; it is a story of cultural and religious beliefs guiding the evolution of a technology. By considering the roles of resistance, reconstruction and rhetoric in the development and launch of the kosher phone we can see how the cell phone in this instance became a religiously cultured technology.

The Religious Culturing of Technology and the Kosher Phone

Resistance

Resistance involves a user community entering into a shared dialogue to evaluate how a technology is influencing or may influence the community. Through dialogue, various negative functions or by-products of the technology may be highlighted which are perceived by the group as needing to be resisted. This process of resistance calls members to reflect on core community values and standards of practice. Through discussing and evaluating the technology, the community reaffirms its values and beliefs and can thus offer standards for acceptable use. In the case of the kosher phone, the haredi community underwent a distinctive public process of reflecting on and eventually resisting standard cell phone technology. Resistance here involved community discussions over a two-year period amongst religious leaders and members on the potential hazards being introduced by cell phone use, which led to official regulations issued by religious leaders on cell phone use and promotion meant to maintain community boundaries and beliefs about technology.

Months before the kosher phone became a reality, discussions arose amongst community member about the troublesome issues being raised by cell phone use in the community. Public outcry became pronounced in early 2004 through several articles and editorial in various haredi newspapers. An article in *Yated Ne'Eman*, a haredi newspaper in Israel, reported that several ultra-Orthodox leaders were calling for the partial ban of cell phones and the blocking of phone signals in religious communities (Tzippori, 2004). Concerns were voiced about the amount of time community members were spending on their cell phones and the financial strain that this was putting on some families. In some discussions the cell phone was described as a 'dangerous weapon' attacking the morality of the community and thus should be battled against. The cell phone was seen as a conduit allowing unacceptable content to find its way into the community.

> Resources are being channelled to sound the warning for cell phones which, unlike computers, follow us wherever we go, and even kosher channels are liable to turn

> into spiritual disaster through an incidental flashing message or another. (Tzippori, 2004, p. 15)

It was not only the religious leaders who were calling for increased regulations. In an article 'Put up the phoneblocks!' a religious journalist suggested the idea of 'charedi-only cell phones' that would be specially marked as a way to supervise users and separate them from unapproved content. The proposed phone would also presumably indicate religious commitment: 'Show me you cell phone and I'll tell you what you're made of. Whether a young man is from our ranks, or whether he is connected to the internet will be apparent' (Tzippori, 2004, p. 15). Thus the idea of the kosher phone provided a way for the community to monitor religious devotion in that ownership could be seen as a way to affirm community affiliation.

According to a report in the Orthodox newsletter *Dei'ah veDibur*, the birth of the kosher phone began when a special committee of ultra-Orthodox rabbis was formed in 2004. The committee's focus was to address concerns raised within segments of the community about the 'spiritual and educational pitfalls resulting from content services accessible using most cell phones' (Kahn, 2005). This committee represented a network of select ultra-Orthodox rabbis, formed specifically to evaluate and make official community policy on cell phone use. Members of the committee met regularly over a period of several months to discuss issues such as SMS content and phone access. One of the committee's first actions was to instruct haredi newspapers to stop carrying all advertisements from cell phone companies in 2004, until an official course of action was agreed upon. This demonstrated the committee's influence within the community.

According to press accounts, the committee met and consulted various technology experts throughout the country on this issue in 2004 and early 2005. The committee was a network of religious authorities and rabbis representing a strong hierarchical structure (Kahn, 2005). Their discussion centred on how the cell phone was influenced by both the economic aspects of the community's life and social practices. 'Setting a fence around Torah' is a Jewish phrase used to describe the setting of strict limits in relation to a new innovation so that its use will not violate Torah law until the device can be fully understood or observed. Thus the council's task was to evaluate the technology in light of the Torah and halacha or religious law. The resulting policies served to clearly tighten the boundaries not only around the technology but also the community itself.

Based on their concerns and needs, and with the help of a Jewish attorney, they entered into negotiations with all of the major cell phone companies in Israel. In the end, MIRS was the only company that 'demonstrated a readiness to meet the haredi's public demands' (Kahn, 2005). The result was that the committee approved and then signed a contract with MIRS authorizing it to produce specially designed 'kosher phones' for community member. Resisting the cell phone began with community reflection on the effect the cell phone was having on the group. This led to the formation of an official structure to evaluate the technology in light of the community's moral and cultural standards, and eventually a call for a change in the technology itself.

Reconstruction

Reconstruction takes place once a technology is recognized as valuable to the user community, but incompatible in its current form. Reconstruction involves discussing what are the acceptable forms or uses and whether any modifications of the technology are necessary in order for it to be integrated into community life. In the case of the kosher phone it meant that the ultra-Orthodox community had to identify what beliefs were challenged, supported or in conflict with the technology. Discussion led to concrete modification to the cell phone and meant restructuring the make-up of technology in order to make it acceptable within community standards.

The haredi leaders began by identifying the values being challenged by cell phone technology. A *Times Online* report stated: 'rabbis say they are not against technology as a matter of course. But they fear that the latest feature packed telephones would provide the opportunity to access corrupting influences' (MacKinnon, 2005). The problematic aspects involved certain features that were associated with secular media-entertainment culture. The ultra-Orthodox have a historic concern that media technology often exposes members to unnecessary temptations through the content they display. Their desire to maintain boundaries between the sacred and the profane creates an apprehension about any technology which might allow images into the home that might lead community members into sin. Their concern for personal purity led to official bans on television within the community, and strict guidelines being issued on acceptable use of the Internet. The cell phone represented another technology that, while providing useful communication capabilities, also promised to introduce unnecessary hazards for those already committed to limiting their exposure to secular content.

Therefore, the deleting of these feature and the shaping of the cell phone simply as a communication tool enabled it to be deemed an acceptable technology. In light of this, MIRS agreed to provide the community phones stripped of all content services, set a block on numbers for phone sex, dating services and other dubious secular offerings ('Israel's "kosher" cell phone testing appetite for growth', 2006). The phone was also embossed with a special seal, and these devices would begin with the same dialling code and prefix. This agreement and adaptation of the technology would make acceptable phones easier to spot. Discounted calling plans offered by MIRS encouraged calling within the newly established network. MIRS also opened three service centres in the key ultra-Orthodox communities of Ashdod, Bnei Brak and Mea Shearim-Jerusalem to market the kosher phone to religious users.

Addressing concerns, such as access to secular content, and implementing features that supported the community, such as creating a distinct community network, reconstructed the cell phone in ways that promoted a specific view of the technology as well as the members' relationship to it. In a statement issued by *gedolei Yisroel* (the chief rabbinical council of ultra-Orthodox rabbis) about the official decisions made by the committee on cell phone use, directives were given instructing those 'who must use cellular devices' to choose only those that have been authorized by the rabbinical committee. According to *Dei'ah veDibur*, the statement read:

> We hereby state our opinion, *daas Torah*, that whoever needs a cellular device (but not those who were instructed not to use any type of cellular device whatsoever, e.g. yeshiva students, Beit Yaakov students and seminary students) may possess and use only this device approved by the above committee and marked with a special imprint. The possession and use of any other types of devices not approved by the rabbinical committee is strictly prohibited, as are advertisements for any unapproved company or device. And let this not be breached and all who heed will be rewarded with pleasantness and will merit *nachas* (happiness) for all of their offspring. (Kahn, 2005)

Before the launch of the kosher phone, discussion about cell phones focused on proper or acceptable use in line with religious beliefs and obligations. However, after the launch emphasis was placed on the model and features of the phone one owned. This was a noticeable shift from critiquing the effects of use to emphasizing the cell phone's form and what it symbolized. The reconstruction of the cell phone by the ultra-Orthodox community involved refocusing the community discussion from the danger of cell phone use to the kosher phone serving as a symbol of communal affiliation and identity.

Rhetoric

Rhetoric refers to identifying the arguments and appeals used by a religious community to frame or interpret a given technology and justify any potential changes made in form or use. While resistance and reconstruction also involved the identification of particular discussions within the community, these categories focus on concrete actions which result from these communal conversations. In the category of rhetoric, focus is placed on identifying how arguments themselves become tools to restructure the presentation and perception of the technology. Here, language employed by the ultra-Orthodox community about the cell phone technology played an important role in framing the technology in a particular light which justified the need to both resist and reconstruct the technology.

The community desired access to the communication capabilities that the cell phone afforded, but wanted to resist exposure to potential secular content. Koshering the cell phone, in practical terms, meant transforming it into a simple communication device devoid of secular content. In rhetorical terms it meant framing the negotiation process in terms of a sacred–secular conflict. As noted in the official statement issued by *gedolei Yisroel*: 'The great spiritual danger inherent in the development of high-tech communications is widely known and we can no longer stand quietly and watch the resulting spiritual decline' (Kahn, 2005).

Instead of describing it as a tension between the community and technology, some community members described this as a battle between religious and mainstream secular culture. This added cultural importance to this negotiation process and affirmed the community's re-shaping of the technology. As an editorial on the launch of the kosher phone stressed:

> Without fanfare, the entire community quietly joined the struggle to protect ourselves from the encroaching tide of filth. Our demand was a simple consumer request: we want new phones that provide basic phone service. We want voice communication and nothing more. ('Opinion & comments: kosher cell phones', 2005)

In some haredi reports, cell phone providers were described as purveyors of secular culture and corrupting influences. By highlighting the fact that the negotiation process was in direct conflict with the secular values of the service providers created a juxtaposition allowing the community to frame the push for the kosher phone in terms of a battle with secular culture. Thus it was not just a demand for special treatment but a moral imperative to pursue and reconstruct this technology.

> All too often the modern world is proving to be *chachomim lehora*—smart people who use their talents destructively. It is up to us to defend ourselves and to take only what is truly beneficial—until the day comes when the whole world is filled with the knowledge of Hashem. ('Opinion & comments: kosher cell phones', 2005).

Their emphasis on community values also indicates that the haredi employed a particular view of technology. Technology should be a tool that helps, or at least does not hinder, one from living a kosher life. Thus when the technology challenges the moral life of the community something about that technology must change. In order to maintain a distance from the secular values seemingly promoted by media culture and other cellular providers, the community had a mandate, with the help of MIRS, to create a protected channel of communication. This rhetoric of the sacred–secular conflict served as a powerful tool to create community cohesion and policy surrounding the kosher cell phone. Though take-up of the kosher phone has been slow—only about 30,000 kosher phones had been purchased by the 600,000-strong Haredi public by July 2006—religious leaders continue to stress its importance as the phone of choice for religious users (Ilan, 2006).

Summary

The emergence of the kosher cell phone and Amish engagement with the phone demonstrate that religious communities may undergo a sophisticated and fascinating process of negotiation related to technology. This article has attempted to lay out how one might study and interpret the interaction of religious groups with technology. First, it means recognizing that religious communities, like other social groups, evaluate, monitor and control their members' use of technology in deliberate ways. Second, it is important to recognize that the 'meanings of technologies, old and new, are culturally instructed and negotiated in the service of particular needs' (Zimmerman-Umble, 1992, p. 192). Thus a religious group's response to a technology cannot be separated from those values that guide all aspects of community life. Finally, detailed reflection is needed to consider the 'dynamic grounds for rejection ...

communication technologies can be as useful as understanding those who accept them' (Zimmerman-Umble, 1992, p. 192). Identifying the process by which religious groups evaluate a technology's potential benefits or hindrances is important because it helps uncover the group's patterns of moral life. This enables us to identify the 'moral economy' of the community and helps to predict how the community may culture these artefacts. The kosher phone highlights how religious culture and belief can serve as an important factor in technological innovation. Other Israeli companies, such as Pelephone and Cellcom, have responded to these developments and in 2006 launched their own versions of the kosher phone (Kalian, 2006). Thus religion, instead of leading communities away from a given technology, may compel them to culture the technology, thereby introducing new features, designs or forms of use that influence industry as well as cultural trends.

References

'A spiritual connection' (2005) The Economist.com, 10 Mar., Available at: http://www.economist.com/science/tq/displayStory.cfm?story_id = 3713855 (accessed 24 Mar. 2005).

Barzilai-Nahon, K. & Barzilai, G. (2005) 'Cultured technology: Internet & religious fundamentalism', *The Information Society*, vol. 21, no. 1, pp. 25–40.

Biddlecombe, E. (2004) 'Cell phone users are finding God', *Wired*, 19 Aug., Available at: http://www.wired.com/news/ (accessed 24 Apr. 2005).

'Company markets kosher cell phone' (2006) *Israel's High Tech Investment Report*, May, Available at: http://72.14.221.104/search?q = cache:Yj7zVDZ9-CcJ:www.ishitech.co.il/0506.pdf + MIRS + negotiations + with + Ultra + Orthodox&hl = en&ct = clnk&cd = 4 (accessed 24 Apr. 2005).

Heilman, S. C. & Friedman, M. (1991) *The Haredim in Israel*, American Jewish Committee, New York.

'Hindus make their offerings by SMS' (2003) Textually.org, 3 Sept., Available at: www.textually.org/archieves/001803.html (accessed 24 Apr. 2005).

Ilan, S. (2006) 'Following the herd', *Haaretz Online*, 3 Jul., Available at: http://www.haaretz.com/hasen/spages/730791.html (accessed 12 Apr. 2005).

'Israel's "kosher" cell phone testing appetite for growth' (2006) *The Star Online*, 31 Mar., Available at: http://thestar.com.my/news/story.asp?file = /2006/3/31/apworld/20060331112814&sec = apworld (accessed 24 Apr. 2005).

Kahn, B. (2005) 'New "kosher" cell phones free of content services', *Dei'ah veDibur: Information & Insight*, 25 Mar., Available at: http://charedi.shemayisraek.com/VRK65amirs.htm (accessed 7 Mar. 2005).

Kalian, G. (2006) 'Company succumbs to haredi pressure, decides to launch special cellular phone for ultra-orthodox users', Ynet.com, 20 Feb., Available at: http://www.ynetnews.com/articles/0,7340,L-3218809,00.html (accessed 10 Mar. 2006).

'Kosher phones for ultra-Orthodox Jews' (2005) Mobliledia.com, 7 Mar., Available at: http://www.mobliedia.com/news/27118.html (2005). (accessed 21 Mar. 2005).

MacKinnon, I. (2005) 'Kosher phone taps into new market for mobiles', *Times Online*, 3 Mar., Available at: http://www.timesonline.co.uk/article0,251-1508115,00.html (accessed 24 Apr. 2005).

'Opinion and comments: Kosher cell phones' (2005) *Dei'ah veDibur: Information and Insight*, 16 Mar., Available at: http://charedi.shemayisraele.com/VRK65ocellphn3.htm (accessed 21 Mar. 2005).

'Phones have been blessed in ritual at a Matsu temple' (2004) Textually.org, 1 Jul., Available at: http://www.textually.org (accessed 20 Mar. 2005).

'Pope spreads the word by SMS' (2003) *Ananova*, 14 Jan., Available at: http://www.ananova.com/news/story/sm_739682.html (accessed 20 Mar. 2005).

Qusti, R. (2003) 'Ramadan: Saudi society under the influence of capitalist culture', *Al Jazeerah*, 29 Oct., Available at: http://www.aljazeerah.info/Opinion%20editorials/2003%20Opinion%20Editorials/October/29%200/Ramadan%20Saudi%20Society%20Under%20Influence%20of%20Capitlaist%20Culture%20Raid%20Qysti.htm (accessed 24 Mar. 2005).

Rheingold, H. (1999) 'Look who's talking', *Wired*, vol. 7, no. 1, Available at: http://www.wired.com/wired/archieve7.01/amish_pr.html (accessed 24 Apr. 2005).

Silverstone, R., Hirsch, E. & Morley, D. (1992) 'Information and communication technologies and the moral economy of the household', in *Consuming Technologies: Media and Information in Domestic Spaces*, eds R. Silverstone & E. Hirsch, Routledge, London, pp. 15–29.

Stadler, N. (2002) 'Is profane work and obstacle for salvation? The case of the ultra Orthodox (Haredi) Jews in contemporary Israel', *Sociology of Religion*, vol. 64, no. 4, pp. 455–474.

Stadler, N. (2005) 'Fundamentalism', in *Modern Judaism: an Oxford Guide*, eds N. de Lange & M. Freud-Kandel, Oxford University Press, Oxford and New York, pp. 216–227.

Tzippori, R. (2004) 'Put up the phoneblocks!', *Yated Ne'Eman*, 15 Sivan 5764/4 Jun., pp. 14–16

'Virgin mobile users confess their sin' (2004) Textually.org, 6 Dec., Available at: http://www.textually.org/textually/archieves/006286.htm (accessed 20 Mar. 2005).

Zimmerman-Umble, D. (1992) 'The Amish and the telephone: resistance and reconstruction', in *Consuming Technologies: Media and Information in Domestic Spaces*, eds R. Silverstone & E. Hirsch, Routledge, London, pp. 183–194.

Pocket Technospaces: the Bodily Incorporation of Mobile Media

Ingrid Richardson

Handheld media and communications technologies are becoming increasingly composite interfaces, combining the functionality of standard telephony, text-based interaction, e-mails and Internet browsing, digital video cameras, PDAs, MP3 players, and game consoles. As devices such as these penetrate and transform everyday cultural practices and spaces, they are effectively transforming the relation between body and world, ready-to-hand and telepresent interaction, and actual and virtual environments. This article focuses on the spatial, perceptual and ontic effects of mobile devices as nascent new media forms, with particular attention paid to the use of games and media content in devices which include telephony as part of their functionality. I will examine both how mobiles are used as phone–game hybrids, and how they also work not simply as communicative conduits but also as 'handy' or pocket containers of data, media content, photo archives and secure microworlds. To date there has been limited attention paid to the corporeality of mobile phones as itinerant game and/or media devices, or to the phenomenological impact of physical mobility on game play and new media consumption/deployment, and the particular sense of transmediatic space and perceptual dispersion that they generate. With these issues in mind, I will consider the impact of recent mobile phone technologies, and share some of the insights afforded by a small ethno-phenomenology of mobile phone and media use in urban Western Australia that is currently in progress.

Phenomenology of Technology: a Word on Method

In its phenomenological focus, my approach is framed within the broad premise that every human–technology relation is also a body–tool relation, and as such every mobile–body merger invokes certain kinds of being-in-the-world, and particular ways of knowing and making that world. I suggest that mobile media usage is quite literally a medium-specific mode of embodiment, a way of 'having a body' that demands a complex socio-somatic adaptation. Indeed, our use of mobile and wearable media can be described in Drew Leder's terms as an ongoing *incorporation* by which we reshape the ability structure of our bodies (Leder, 1990). In these terms, then, how does this portability and wearability—the prosthetic and orthotic capacities of mobile media—impact upon the body–technology relation? How can we critically interpret the particular *technosoma* to emerge from this relation? Do our perceptions of media and game space become transformed by the physical mobility enabled by portable devices?

Merleau-Ponty's (1962) emphasis on our corporeal and perceptual engagement with the environment can provide valuable insights into the interpellation of bodies and tools in all human–technology relations; in particular, his notion of the *corporeal schema* or *body image* can effectively be applied to the relationship between mobile media and embodiment. The corporeal schema describes the 'expandable', inherently open and tractable nature of embodiment. In the context of everyday activities, the experience of one's own corporeal schema is not fixed, but may encompass a range of potential body-aspects and body-images in the form of material and technological mediations, and cultural and historical contexts. For example, learning to drive a car involves learning the spatial organization and limits of the vehicle, its speed, the hand–wheel–direction vector and foot–brake–deceleration continuum, and so on. Within the material shape and capacities of the car, we adjust our physical deportment, spatial orientation, and our entire physical relationship with the world. Drivers need to train their car-body and accommodate entirely new ways of thinking about and moving through space: becoming familiar with emergent ratios of hand–eye and foot–eye co-ordination, judging distances with the visual device of the rear-view mirror (a new vision which warns 'objects are closer than they appear'), and many other techno-corporeal proficiencies. In addition, we must acquire a very complex somatic and semiotic literacy of the built environment, including road width, signage and artificial lighting. Initially, learning to drive involves constant attention and concentration, because we must consciously orient our bodies towards the unfamiliar spatial and motile logic of the car, but after some practice driving becomes habitual, and our corporeal schema properly 'dilated' to accommodate the tool. In this way, the car becomes an aspect of our embodiment, part of our repertoire of proprioceptive skills; we have appropriated the body of the car as a temporary body or quasi-body, a supplemental being-in-the-world. Indeed, it is the essential mutability of our *soma*, which adapts according to a complex range of cultural, personal, physical and medium-specificities, that describes the very nature of embodiment.

Both *as* and *in* context, our embodiment exists as a complex interspersion of physicality and biology, material and cultural environment, somatic memory and habit. Within this relational ontology qua embodiment and technology, the body is a material-semiotic assemblage with constantly shifting boundaries; but also, in my analysis, as quite literally *mediatropic*—disposed both metaphorically and materially towards media technologies. Thus, for example, we are now corporeally and conceptually familiar with the phenomenon and experience of telepresence—perception and communication at-a-distance—in everyday life. As Eugénie Shinkle (2003) suggests, media technologies institute 'material parameters', proportions of attention and inattention, by which we measure varying degrees of 'perceptual reach' from objects and others in the world. As is most often the case with our use of contemporary media, for example, the hand–eye–screen interface or the hand–eye–remote control arrangement works as the preferred default. Yet as I will suggest, our use of handheld screens when we are on the move complexifies the body–tool corporeal schema particular to screen and televisual media.

Our engagement with screens at a perceptual and phenomenological level is, of course, deeply embedded in an assorted history of image technologies and collective media–body interfaces. In McLuhan's (1964) understanding of medium-specificity, each communication medium works to 'fix' particular sensory ratios, stipulating forms of knowledge and orchestrating the very structure of perception. Yet the body–media relation specific to screen and micro-screen interfaces is not adequately achieved simply via an application of McLuhan's notion of sensory ratios. The corporeal schema *exceeds* a purely sensory perception both in its intercorporeal mingling with the world and in its visceral dimension. Thus rather than use the idea of 'sensory ratio' I will explore the mobile–body relation in terms of *somatic involvement*, a concept which recognizes the medium-specificity of both our sensorial spectrum and more broadly our corporeal schema in relation to technologies in use. Mobile media technologies, and tele-technologies more generally, are therefore not simply prostheses or augmentations of our sensorium, but tools which impact upon our body limits, shifting the variable boundaries of embodiment, and altering our sense of having a body: they educe altered 'involvements' of the soma. In what follows I will focus on the media–body relation in terms of the specificity of micro-screens, and the spatial ontologies and corporeal schematics particular to our use of mobile phone media.

A Phenomenology of Mobile Screens

Contemporary Western culture can be said to have a particular epistemological and perceptual bias, an ocularcentrism which works to prioritize visual and screen representations. Critics such as Weibel go so far as to say that

> The primacy of the eye ... as the dominant sense organ of the twentieth century is a partial effect of a technical revolution that put an enormous apparatus to the service of vision. The rise of the eye is rooted in the fact that all of its aspects (creation, transmission, reception) were supported by analog and digital machines. The triumph of the visual in the twentieth century is the triumph of techno-vision. (Weibel, 1996, p. 339)

It has been argued that the prevalence of vision-metaphors and visualizing technologies have sustained our epistemological habit of collapsing seeing with knowing, at the same time separating subject (the seer) from object (the seen), and elevating sight as the one 'true' sense enabling disembodied and objective knowledge.

Theorists such as Romanyshyn (1989) have claimed that screen technologies in particular, by providing a frame or window to organize and contain what can be seen, have continued this legacy of visualism. Yet while on the one hand it might seem that new media screen technologies of today are wholly consonant with the ocularcentric paradigm—as Heidegger (1977) claimed, gathering up and ordering the world 'as picture'—I would argue that mobile media screens set up an entirely different relationship to embodied perception, and require a corporeal schema quite at odds with our usual habits pertaining to screens.

In order to grasp the epistemic, ontic and phenomenological status of screen media, it is important to trace their ocularcentric legacy; by understanding this history we can then interpret how mobile screens in particular work to bewilder classical notions of visual perception, agency and knowing, and how they have contributed to what Don Ihde refers to as contemporary polymorphic or compound vision:

> Ours is a pluricultural vision ... It is a series of multi- and alternative visions, symbolized in the growing presence of the multivisual screens we have become familiar with ... One scans the multiple screens, focusing here, then there and, out of the mélange, forming new directions and possibilities. (Ihde, 1993, p. 29)

The multistable and disparate nature of contemporary vision is thus the partial effect of the many screens encountered in the everyday—televisual, cinematic, information/text display, closed-circuit, video—each with their own technical, environmental and interfacial specificities. What we experience is an aggregate vision, a continuous slippage and merging between televisual events, spatio-temporal zones, and genres of visual meaning (photorealism, animation, simulation). Concurrently, we develop a literacy of a range of imaging techniques, of various modes of visual and audio intervention, such as the freeze-frame, the close-up, and slow motion, and perhaps even an awareness of some hidden qualities of digital imaging including manipulation, repeatability, and compression. The work of Lev Manovich, Peter Lunenfeld and others on new media and 'interface culture' articulates some of the visual specificities of digital and screen media, from the multiple layering and spatial incoherence of the Windows interface (Manovich, 2001), to the visual manoeuvres of hypertext and virtual reality (Lunenfeld, 1999).

In a phenomenological context, we can also determine some general properties of what Introna & Ilharco (2004) call 'screen-ness'. In contemporary life, screens are often a primary focus of our attention and concern: they literally display that which is relevant or worthy of notice. This property of relevance has little to do with the specific content of any particular screen display; rather, it indicates:

> a particular involvement in-the-world in which we dwell and within which screens come to be screens. It is not up to anyone of us to decide on the already presumed relevance of screens; that is what a screen is—a framing of relevance, a call for attention, a making apparent of a way of living. (Introna & Ilharco, 2004, p. 227)

Yet while it is possible to describe the broad-spectrum nature of screens in this way, the specific phenomenology of screens—in terms of their functionality, size, sensory and somatic involvement, patterns of use and behaviour—must also be accounted for. As Chris Chesher points out, ways of seeing and 'conventions of looking' are not innate or given, but culturally and materially contextual, such that each 'new visual technology emerges with its own conventions—its own structures of feeling' and concomitant mechanisms of attraction and counter-distraction (Chesher, 2004). The mobile phone is clearly emerging as a distinctive multi-modal screen in its own right, with its own perceptual ratios and techniques.

Describing the particular 'screen-ness' of mobile phones must also involve an account of how the mobile is not just, or even primarily, a screen; it enacts both separately and combined visual, haptic and acoustic incursions into our corporeal schema, and demands variable and oscillating modes of somatic involvement. At times the screen is denied in favour of voice activation and communication via Bluetooth technology, but often the finely attuned coupling of hand and mobile device, the haptic familiarity of the keys, and the ability to use the phone one-handed while texting or navigating with the thumb, also vitiate against focused screen engagement. In part (one-handed) haptic and aural use allows relatively uninterrupted engagement and communication on the move, and is thus preferred in some contexts and situations. Newer phones with PDA functionality often have no actual keypad, but rather include a stylus and touch screen for navigation and 'dialling'. This interface requires two hands, and the use of the stylus rather than forefinger is necessary for finer navigation and texting/typing. In focus group interviews, a number of what I would call 'expert users' of mobile phones, who regularly upgrade to just-released models and utilize most or all the functions in their devices, commented on their experience of touch screens and styluses. Interestingly, although several of the expert users desired or owned an all-in-one device combining a high-quality digital camera, PDA, MP3 player and mobile phone, they also claimed that while the stylus interface was well suited to PDAs (which required more focused attention and thus warranted stopping or sitting and two-handed usability), mobile phone functionality (including texting and navigating) should be optionally one-handed.

In part this distinction between functions (if not devices) seems a matter of habit, and the counter-habitual nature of stylus use in relation to phone functionality. As Robertson (2005) points out, habit, as the 'sedimentation' of action, rearranges our corporeal schema and so quite literally reorganizes agency in the world. With sedimented use of a particular mobile phone, 'through an ongoing adjustment of motility', we take 'the motor space of our interaction' *into* our hands and the space of our bodies, such that we come to 'know' its model-specific characteristics in the same way that we know the placement of our own limbs and fingers. She writes:

> Merleau-Ponty maintained that the process of learning to type quite literally incorporates the space of the keyboard into bodily space, the situated space of our phenomenological bodies. The space of our cars ... of our pens, of our cameras, of our PDAs and of our mobile phones is incorporated into bodily space in the same way. (Robertson, 2005)

Other 'everyday' or non-expert users of mobile phones, in separate interviews, stated emphatically that they would never use either a stylus or Bluetooth device, but had difficulty providing a reason except in the vaguest of terms ('I don't like them'). A possible explanation may be that both stylus and Bluetooth earpiece affect a less intimate relationship between the hand the phone, and there is no doubt that phenomenologically, hand–tool relations are one of the most significant for our corporeal schematics. As Amparo Lasen insightfully comments, people will often hold and 'handle' their phones even when not in use:

> The way mobile phones are held and touched is one of the aspects that make this relationship different to other ICT devices. The attachment to mobile phones is revealed by the transformation from being an object always at hand to being almost always in the hand and close to the body. (Lasen, 2004)

In becoming an incorporative aspect of the hand, the mobile phone thus enters into an intimate and habitual relationship with a body part that is in itself of some consequence as a communicative and world-shaping tool.

Introna and Ilharco suggest that screens of all kinds enter our involvement-in-the-world at the moment we turn them on, at which point we reposition our attention and 'sit down, quit—physically or cognitively—other activities we may have been performing, and watch the screen' (Introna & Ilharco, 2004, p. 225). Yet this 'frontal' relationship which is typical of our engagement with most screens—where the mediums of cinema, television and computer can be said to discipline the body more or less into a face-to-face interaction—is thoroughly challenged by the mobile media screen. Our interaction with mobile screens is rarely marked by such dedicated attentiveness; indeed, our 'turning towards' them is usually momentary (checking for a text or missed call) or at most can be measured in minutes. In fact, even in the seemingly committed practice of game-play, mobile phone engagement is characterized by interruption, and sporadic or split attention in the midst of other activities, a behaviour quite distinct even from handheld console game-play on the Nintendo DualScreen (DS) or PlayStation Portable (PSP). This is recognized by the growing mobile phone game industry and its labelling of a key market as 'casual gamers', who play at most for five minutes at a time and at irregular intervals; it seems mobile phone gamers don't want immersion (Hume, 2005). In interviews, several expert users somewhat paradoxically claimed both that they 'had no time' to play games on their mobile phones because it required continuous visual attention, yet frequently and often on a daily basis *did* play games when they had a few minutes to spare (while waiting for an appointment, on public transport, in a queue).

The suggestion that the proliferation of computer interfaces and digital games have us fixated on screens, and for the most part immobilized by the continuous frontal demands of the interface, is clearly problematized by our use of mobile micro-screens and the phenomenology of the casual gamer. In his analysis of the console game, Chesher (2004) distinguishes between ways of looking specific to cinema, television and console games—characterized by the gaze, the glance and the glaze respectively—yet the ill fit of mobile phone conventions of viewing within this schema perhaps highlights the need to theorize a cross-modal 'regime of vision' specific to both game and non-game use of portable micro-screens. In fact, the mobile phone device crosses over each of these ways of looking if only because we can—and do—watch movies and live TV, and play games, on our phones at varying levels of immersion and distraction. Chesher suggests that console games are 'sticky', holding the player to the screen via both a quasi-visceral immersion in depth-perspective virtual space, and a haptic attachment to the hand-controller and peripherals: 'In glazespace ... players suspend their awareness of their day-to-day world to become cybernetically suspended within a virtualised sensorimotor space of the game world' (Chesher, 2004). Casual gamers, on the other hand, must deliberately avoid this 'stickiness' so that they are perpetually ready to resume their temporarily interrupted activities. Mobile game developers are also attending to the specificities of the mobile phone in designing games specifically for the device, explicitly recognizing what new game opportunities handsets allow—namely mobility, connectivity, camera functionality and GPS capability. Some location-based games (LBGs) such as Mogi (beta-tested successfully in Japan since 2003), integrate virtual objects into the camera view of the actual environment, and require teams of players to 'capture' them at leisure, or whenever a player happens to be walking past a particular location in the city (Hall, 2004a). Social designer of networked environments Amy Jo Kim suggests that Mogi is ideally suited to the casual gamer: 'It nestles in your everyday life, rather than requiring you to change your behavior ... It amplifies your ordinary behavior—it changes going on an errand into a piece of a game' (Hall, 2004a). Such 'mixed-reality' games or next-generation mobile entertainment platforms, rather than creating an escape from 'real life' through screen immersion, work to integrate play and game interaction into the patterns of everyday life and work.

Digital video and photograph capability via the mobile phone screen is also demanding a perceptual and cultural literacy particular to the medium. More so than the digital camera, the mobile phone is customarily accepted almost as a body-part or appendage, and along with this ever-presence, its multifunctionality renders its status as 'camera' ambiguous. In this way the mobile camera educes a particular kind of ubiquitous visual access, a photo-readiness enabling the capture and containment of immediate and often intimate objects and events. As Pescovitz describes, the photographs exhibited at the Mobile Phone Photo Show (http://www.mpps.tv) in San Francisco in 2004 showed how the intimate relationship we have with mobile phones—more intimate perhaps than with any previous media or communications technology—inevitably 'bled' into the images (Pescovitz, 2004). Mobile phone theorists Daisuke Okabe and Mizuko Ito (2005) argue that because the camera phone embodies the phenomenological characteristics of the mobile it becomes a device of different functionality and expectation. They suggest that the mobile phone has three central properties: it is pedestrian (on-the-street, pervading all settings and locations), portable (on-the-body, both inside and outside the home setting), and personal (literally both a self-portal and private archive). Appropriating these properties, the phone camera's ubiquitous visual access effectively heightens our visual awareness of the everyday, converting every situation into a potential photo or mini-video narrative opportunity (Okabe & Ito, 2005). Interestingly, in discussing their use of the camera phone, several interviewees in my study claimed that although they kept many photos in their phone's memory (often over 100) they never or very rarely sent mms messages or photos to friends, preferring to share their screens in face-to-face interactions or keep them as private reminders of places and people. This use of the mobile phone, as a portable, personal, safe and always-accessible data archive carried on the body—one of the central properties identified by Okabe and Ito—indicates how the mobile phone is so handily appropriated as a pocket technospace or container, becoming integral to our spatial and corporeal schemas in contemporary culture.

Pocket Microworlds and Mobile Gamespace

There is no doubt that televisual technologies have irretrievably altered our sense of embodied 'location' and 'presence'. In twenty-first-century teleculture it is no longer possible to consider space in terms of the dichotomized categories of here/there, near/far, personal/private, inner/outer or presence/absence, dialectics which dominated our understanding at the beginning of the twentieth century. Technological developments ranging from the telephone through to radio, television, cinema, and video games have created quasi-spaces where a sense of presence can be felt beyond the location of the physical body. Both established and new media technologies frequently function by appropriating space as a framing metaphor to enable consumption and use, and there is much to be said about the configuration of technospaces and media spaces in their specificity, and the relation between these spaces and their effect on our corporeal schematics. In what follows I will consider both the interior and connected spatiality particular to the mobile phone.

In the traditions of Western philosophy, metaphors of space are often formulated as particular kinds of containment, such that our being-in-the-world is always also a mode of 'holding' and 'having' (Hefferon, 2002, p. 12). More recently, the representation of computers, game consoles and portable media devices as spaces of containment or microworld reservoirs implicitly relies on the already recognizable container-like properties of media apparatuses such as the television and radio. As Sofia observes, within the average home:

> Books, photograph albums, telephone directories, the television, the stereo, cassettes and CDs: all these media technologies ... [have] their container-like aspects. [E]lectronic and print media are storage technologies for other spaces and experiences: a CD or tape can open up a whole concert, or an aural landscape of feelings; a book can disclose another world. (Sofia, 2000, pp. 189–190)

Indeed, televisual, computational, and game spaces are dominated by this metaphor of containment, and our ability to enter or *be* in these spaces—and to hold and carry them around on our bodies—is predicated on a perceptual and corporeal assimilation of this metaphor. The Motorola Razr advertisement broadcast in 2005 very effectively captured our collective understanding of the mobile phone as a device of containment. The ad depicted a woman seated alone in a domestic loungeroom in which there were several state-of-the-art digital interfaces such as a laptop and flat-screen panel; the walls and furniture fold into each other, and finally they and 'the rest of the world' collapse into an open flip-top Razr which comes to rest on the woman's hand, demonstrating both the safe and handy containment of her environment, and the simultaneous control she has over both actual and data worlds. We both desire and know the impossibility of achieving a neat, compact and foldable being-in-the-world, yet this 'as if' sense of containment is a common experience of mobile phone users; interviewees in my study frequently referred to their phones as microcosms of their lives, far exceeding the containment capacities of wallets and handbags.

Current developments in location-based gaming via the mobile phone make possible even more radical changes to the way we experience the relation between our corporeality and actual and virtual space. Michel de Certeau has famously argued that 'a spatial order organizes an ensemble of possibilities', such as pathways and places where one can go, or objects blocking or redirecting one's path; at the same time the pedestrian trajectory actualizes and creates some of these possibilities simply through the 'improvisation of walking' (De Certeau, 1984, p. 98). The players of Mogi and other LBGs quite often alter their trajectories through the city, dynamically reworking the spatial order and 'ensemble of possibilities'. A frequent player of the game describes how his trips to the city have become physically 'randomized' or diverted as an effect of the game, such that he gets 'a chance to discover part of the city that I ignored, [motivated] to check out that parallel street I never took' (Hall, 2004a). Location-based and next-generation mobile phone games may potentially work to seamlessly integrate the corporeal schematics of actual and virtual worlds as they are actively negotiated on the move. LBG creator John Paul Bichard imagines a scenario where the physical environment becomes a 'vast game engine', and objects, places and people 'part of an intertwining series of episodes' (cited in Frauenfelder, 2005). Similarly, Justin Hall envisages a near future where 'virtual data objects are scattered all over the real, physical world' and mobile phones have tangible context-sensitivity (Hall, 2004b).

In location-based mobile phone gaming the mobility of the physical body becomes key to game-play, and the mobile phone and body act together as feedback mechanism and avatar within the combined actual–virtual space (Lahti, 2003). The body is not represented in the game (as in, for example, the point-of-view or over-the-shoulder style of console and computer games), it is *in* the game, and the game *in* the world, enacting a seamless continuity between the virtual and the physical, and conflating the vicarious link between body and avatar. Moreover, as theorists and developers such as Terry Rueb (2004) and Justin Hall (2004a) suggest, location-based mobile games have the potential to engage the pedestrian and motile body and habitualize a new range of gestures and movements that bypass the current room- or seat-based restrictions of digital gaming. The elements which would combine to create such next-generation environments are not simply imaginary and fictional with microworldly integrity, then, but mutually contingent negotiations between actual and virtual domains—that is, not only can virtual objects be implanted into actual environments, but one can more radically envisage that widespread use of such platforms may eventually effect changes to architectural design and the planning of urban spaces. As such, we might see emergent spatial ontologies of a kind never before experienced in such a collective and interactive fashion.

Conclusion

Within contemporary theory concerning televisual and screen media, experience is frequently described as disembodied and/or predominantly visual. Against these interpretations, phenomenology can consider the ways in which tele-technologies modify the body, and the kinds of embodiments afforded by telepresent and mobile media. In this article I have discussed a number of phenomenological issues relating to second- and third-generation mobiles, including their unique screen properties, the emergent screen–body ontology specific to mobile phone media, their capacity to reconfigure our relation to actual and virtual space, and their particular impact upon our experience as-bodies. As I have suggested, mobile devices antagonize any notion of a disembodied telepresence that is sometimes seen as endemic to screen-based media. While recognizing the 'distancings' or alterations to somatic involvement that may inhere in mobile media and communication, mobile phone 'being' is nevertheless very much embodied, motile and *in*-the-world. The mobile device is simultaneously—and often equally—an aural, visual and haptic interface, requiring an evolving sensibility and inter-modal literacy of which our corporeal schemas, in their ontic flexibility, are inherently and essentially capable.

References

Chesher, C. (2004) 'Neither gaze nor glance, but glaze: relating to console game screens', *SCAN: Journal of Media Arts Culture*, vol. 1, no. 1, Available at: http://scan.net.au/scan/journal/display.php?journal_id = 19 (accessed 8 Apr. 2005).

De Certeau, M. (1984) *The Practice of Everyday Life*, University of California Press, Berkeley.

Frauenfelder, M. (2005) 'Are we there yet?', *The Feature*, 2 Mar., Available at: http://www.thefeaturearchives.com/topic/Gaming/Are_We_There_Yet.html (accessed 19 Feb. 2006).

Hall, J. (2004a) 'Mogi: second generation location-based gaming', *The Feature*, 1 Apr., Available at: http://www.thefeaturearchives.com/topic/Gaming/Mobi_Second_Generation _Location-Based_Gaming.html (accessed 19 Feb. 2006).

Hall, J. (2004b) 'Mobile entertainment: the power of play', *The Feature*, 16 Jun., Available at: http://www.thefeaturearchives.com/topic/Gaming/Mobile_Entertainment__The_Power_of_Play.html (accessed 19 Feb. 2006).

Hefferon, E. (2002) *Container–Agent–Space*, Honours Thesis, Murdoch University, Western Australia.

Heidegger, M. (1977) *The Question Concerning Technology and Other Essays*, Garland, New York.

Hume, T. (2005) 'Casual gaming', *The Feature*, 8 Feb., Available at: http://www.thefeaturearchives.com/topic/Gaming/Casual_Gaming.html (accessed 19 Feb. 2006).

Ihde, D. (1993) *Postphenomenology: Essays in the Postmodern Context*, Northwestern University Press, Evanston, IL.

Introna, L. D. & Ilharco, F. M. (2004) 'The ontological screening of contemporary life: a phenomenological analysis of screens', *European Journal of Information Systems*, vol. 13, pp. 221–234.

Lahti, M. (2003) 'As we become machines: corporealized pleasures in video games', in *The Video Game Theory Reader*, eds M. J. P. Wolf & B. Perron, Routledge, New York and London.

Lasen, A. (2004) 'Affective technologies—emotions and mobile phones', *receiver*, no. 11, Available at: http://www.receiver.vodafone.com/11/ (accessed 14 Nov. 2005).

Leder, D. (1990) *The Absent Body*, University of Chicago Press, Chicago.

Lunenfeld, P. (1999) *The Digital Dialectic: New Essays on New Media*, MIT Press, Cambridge, MA.

Manovich, L. (2001) *The Language of New Media*, MIT Press, Cambridge, MA.

McLuhan, M. (1964) *Understanding Media: the Extensions of Man*, Routledge, London.

Merleau-Ponty, M. (1962) *Phenomenology of Perception*, Routledge & Kegan Paul, London.

Okabe, D. & Ito, M. (2005) 'Personal, portable, pedestrian images', *receiver*, no. 13, Available at: http://www.receiver.vodafone.com/13/ (accessed 22 Jan. 2006).

Pescovitz, D. (2004) 'Art of the cameraphone', in *The Feature*, 3 May, Available at: http://www.thefeaturearchives.com/topic/Culture/Art_of_the_Camerphone.html (accessed 19 Feb. 2006).

Robertson, T. (2005) *The Stamp of Movement in Human Action*, Working Paper, University of Technology Sydney, Available at: http://research.it.uts.edu.au/idwop/downloads/RobertsonMovement.pdf (accessed 5 Dec. 2005).

Romanyshyn, R.D. (1989) *Technology as Symptom and Dream*, Routledge, London.

Rueb, T. (2004) 'Syncopated space—wireless media shaping human movement and social interaction', *receiver*, no. 10, Available at: http://www.receiver.vodafone.com/10/ (accessed 22 Jan. 2006).

Shinkle, E. (2003) 'Gardens, games and the anamorphic subject: tracing the body in the virtual landscape', *Paper presented at the Digital Arts Conference*, 19–23 May, Melbourne, Australia, Available at: http://hypertext.rmit.edu.au/dac/papers/Shinkle.pdf (accessed 8 Nov. 2004).

Sofia, Z. (2000) 'Container technologies', *Hypatia*, vol. 15, no. 12, pp. 181–201.

Weibel, P. & Druckrey, T (1996) 'The world as interface: toward the construction of context-controlled event-worlds', in *Electronic Culture: Technology and Visual Representation*, ed. T. Druckrey, Aperture, New York.

Becoming the Milky Way: Mobile Phones and Actor Networks at a U2 Concert

Chris Chesher

The world's biggest band, U2, brought their spectacular *Vertigo* show to the Sydney Telstra Stadium in November 2006. Alongside 70,000 people who came to each performance, tens of thousands of mobile phones also attended. Seeing so many phones participating in such an event led me to ask: what connections could be made between the world's most popular band and the world's most popular portable electronic device? How did thousands of individuated connections mediated by mobile phones transform a 'mass' spectacle? What distinctive rituals of mobile phone use might be particular to the stadium? At the same time, what invisible actors were at work in the background to allow these rituals to be performed so effortlessly?

Of all recently developed media technologies, mobile phones have a flexibility that allows them to articulate smoothly into a diversity of anticipated and unanticipated programmes of action. While many of these programmes take up certain phone 'features' and network capabilities, the openness of these devices to semi-improvised uses allows them to adapt to anything from everyday routines to special events. Mobile phones were constantly present in and around the U2 show, but their value and functions were different at different times. They helped some people to physically locate friends, or to connect with those not immediately present. Their cameras created, stored and transmitted images, and their backlights even became part of the light show.

This article looks at how a number of distinctive, but sometimes apparently trivial, uses of mobile phone are structured by the pragmatics of events at the U2 concert. The show was staged in the massive non-place of the stadium (Augé, 1995), which has more in common with other stadiums anywhere than with other places in Sydney or Australia. The site provided a very suitable ethnographic location because the environment and the event presented certain challenges and opportunities for visitors and performers alike, encouraging some innovative uses of mobile phones. Recent ethnographic research has revealed how communities in specific national and regional contexts find their own uses for mobile phones (Donner, 2005; Horst & Miller, 2006). Other studies have explored how techniques for using particular phone features, such as the short message service (SMS), emerge over time (Harper *et al.*, 2005; Taylor & Vincent, 2005). I wanted to see whether patterns of use might be apparent even over a few hours, at a major popular culture event.

Useful in analysing this diversity of mobile-mediated events has been Bruno Latour's 2005 book *Reassembling the Social.* This book presents itself as an introduction to actor network theory (ANT). ANT research typically combines ethnography with analyses of the heterogeneous human and non-human actors at play in given networks. Following this approach, I will analyse the human and non-human actors I observed before and during the performance: the stadium, performers, lights, music, fans, mobile phone handsets and base stations. I was looking particularly for events of translation or transformation, whereby one actor has been substituted for another, or where an actor forms a connection with another, or a network is otherwise rearranged.

Latour connects his work with Gabriel Tarde, a controversial but significant figure in turn-of-the-twentieth-century sociology. Unlike the grander sociological tradition founded by Durkheim, which seeks to discover 'social facts', Tarde proposed an atomic, or monadic conception of change. He argued that changes emerge as innovations that begin at a molecular level and multiply themselves through imitation:

> This is why any social production having some marked characteristics, be it an industrial good, a verse, a formula, a political idea which has appeared one day somewhere in the corner of the brain, dreams like Alexander of conquering the world, tries to multiply itself by thousands and millions of copies in every place where there exists human beings and will never except if it is kept in check by some rival production as ambitious as itself. (Latour, 2005, p. 15)

Sensitized by reading Tarde and Latour, I approached the show with an understanding of the possibility that innovations can emerge from the smallest thoughts, behaviours or actions of human and non-human actors. My observations needed to examine actors and events of all spatiotemporal scales. The most transitory event can be translated into something else. One-time improvisations by users can become collective habits. These habits can inspire new designs. New designs can generate new markets. The transformation of ringtones from functional features to entertainment commodities is only one example in the accidental history of mobile phones (Gopinath, 2005). These changes involved the enrolment of MIDI and RealTone standards, musicians, publishers, and commercial download services, new meanings and uses, and extending the network of mobile phone actors substantially.

The stadium, in spite of its monolithic form, also operates as a complex network of actors. It has to support bodies, equipment, catering, and mobile phone connections, apparently effortlessly. Stadiums are classical architectural actors that mediate the flows of large crowds by enrolling a combination of physical features and human attendants. Traditionally they have accommodated throngs of people, rather than individuals. More recently, emerging technologies, including mobile phones, have afforded increasingly complicated and individuated attachments between stadiums and patrons. As with all complex social situations there is some arbitrariness to what people might do from moment to moment. At any time, any patron might join a queue, buy a T-shirt, make a spontaneous phone call, or do something less expected. Many actors in the stadium address crowd members as part of a mass: security barriers, security guards, jumbo screens and speaker stacks. Other actors address patrons with greater granularity: numbered tickets and seats; stands selling souvenirs, food and drink; and mobile phone base stations. Scanners read and verify the tickets. Ushers show patrons to their seats. Surveillance cameras look for antisocial behaviour. Security guards wait on stand-by for spontaneity of undesirable kinds: terrorist threats, drunkenness, and so on. Other actors seek opportunities to capture the value of impulsiveness, as T-shirt purchases, phone calls or whatever.

The U2 performance itself ideally provides an experience that is apparently total, establishing both a personal connection with each member of the audience, and filling the stadium with a panoramic audiovisual spectacle. Each performance actualizes a virtual relationship between band and fans that has been established through repeated playbacks of songs, telling narratives of band biographies, and the personal histories of each fan. These intangible connections with audiences are bolstered architecturally in the set design with actors such as the 'Ellipse' stage that wraps around the audience in close, and four I-MAG video screens visible from the furthest seats (Cunningham, 2005). The show also makes many other more mundane and distant associations off-stage: a touring crew of 88, with 100 local assistants that sets up and bumps out every performance; and a production office in Dublin. Global chains of actors comprise the production, distribution and promotion of the band's CDs, DVDs, music videos, and ring tones.

Those of us who came to see the show, with or without mobiles, shared some common connections. As ticket holders we passed through the stadium's scanners. We also had some personal identification with the music or affiliation with our peers. We had probably invested significant time and money on consuming recordings and images, learning the songs, and absorbing the milieu. These repetitions had generated some affection and identification with the music. Fandom might also find expression in fan zines and websites, posters on bedroom walls, or U2 iPods in the pocket. These investments would affect the attitude to the event, giving a sense of anticipation.

Becoming a mobile user also marks identity and provides certain capacities for action. The mobile phone opens up hybrid spaces, joining physical spaces in which callers roam physically with the connected spaces where their conversations take place (de Souza e Silva, 2006). The phone as material object must be matched by competence to use it, and sensitivity to the circumstantial forces that call it into action. Beyond the basic call functions users must master at least part of a suite of mediating technologies including a clock, caller ID, games, camera, WAP, Internet, MP3 player, SMS services, reminders, and contact lists. Many mobile users even become emotionally attached to their phones (Vincent, 2005), an attachment that is also the legacy of everyday repetitions of similar, but different, operations over months and years. For the experienced user, a mobile phone provides apparently universal connectivity to social and information networks, and is therefore an indispensable companion for the U2 show.

Dancing into the Stadium

As we approached Telstra Stadium—a colossal building symbolically supported by Australian's dominant telecommunications provider—mobile phones were already conspicuously out of their pouches. The miniature mobile phone was an appropriate answer to the intimidating and gigantic scale of the stadium, which dwarfed the human bodies approaching. At this point I saw several people using mobiles in a distinctive way, partly conquering that space. They adopted a characteristic gesture—neck craned, gazing intently into the distance, pushing the phone into an ear, and moving purposefully around. Their voice was a little too loud as they concentrated on the task of triangulating positions with their friend by making reference to landmarks that both callers could see. This ritual was punctuated by a moment of recognition as the callers saw each other. The phones were lowered simultaneously with a slightly embarrassed smile that recognized the absurd dance they had both just performed. This behaviour, a spatially specific form of micro-coordination (Ling, 2004), is not something that phone users are trained to perform. It's not in the phone manual, or the telecommunications providers' instructions, or the venue's advice. Perhaps in the future this artful dance will be supplanted by some technical actor that transmits global positioning system coordinates. For the moment, the locative dance is something that emerges from the participants' programmes of action—finding each other—the time sensitivity of the event, the movement and impenetrability of the crowd, the alliance with mobile phones, and the scale of the public space.

The freedom for callers to improvise in this way is dependent upon the reliable operation of both phones and networks. While this trustworthiness is mostly taken for granted, it is never assured. A dropout or network overload can demand changes in programmes of action, prompting moments of innovation, even if painful at the time. Our dancers would have to meet at the gate, go to their seats, or find each other in some other way. When signal strengths drop into the red, a battery dies, or a handset is stolen, the phone and the whole assemblage (people, scene and objects) are transformed. However, the stadium features an array of actors seeking to head off such breakdowns. The mobile-wielding multitude can afford to act impulsively in this space because they are embraced by the coverage of an intense concentration of connectivity in and around the stadium. The stadium is a centre that aggregates spectators and facilitates connections of a range of kinds: interpersonal, visual, auditory and telecommunicational. It holds as many as 110,000 people, so requires a massive cellular capacity. When it was built, Stadium Australia, as it was then known, had the densest concentration of mobile phone capacity anywhere in the world (Taggart, 2000). Even if everyone in the stadium made a call at once, the systems could accommodate the load. The stadium and surrounds incorporate over 4,600 kilometres of optical fibre, well over the current required capacity.

The success of the visiting show relied on dependable chains of actors that extend well beyond mobile connectivity, of course. The project director Alan Patching refers to it as a 'fourth generation' facility, meaning 'it is multifunctional and state-of-the-art, including the flexibility to incorporate, with relative ease, the technical and commercial demands of the future' (Patching, 1999, p. 15). Designers attempted to anticipate and put in place an actor network open to future needs and desires of spectators, sponsors, broadcasters and performers.

Messages from Our Seats

Once we found our seats, several hundred metres from the stage, mobile phones started showing their faces again. The main programmes of action seemed to be making contact with people elsewhere in the stadium. The preferred mode of connection was SMS. In the noisy environment of the support band, SMS was the only realistic mode of communication. This mode was sufficient for establishing some sense of where else in the stadium the others were located. Even with the limited messages, the phone briefly extended and transformed the relationship between sender and recipient. Sending and receiving SMS messages filled the temporal gap in the collective anticipation before the show. It offered a way to break down the interpersonal distances at play in such a large venue. It helped to symbolically claim a tightly constrained personal space, inviting friends to visit, even as a ringtone and 160-character bundles. This activity gave people something to do with their hands: now that smoking is not allowed, mobiles phones are the props that allow you to look cool when you're feeling out of place.

It is worth remembering that SMS connections cannot be taken for granted: not because of network capacity, but because of standards and politics. The uptake of SMS in Australia stalled until the providers entered an interconnection agreement in 1999. Until then only subscribers to the same network could send each other messages, so very few did. Without network interconnection agreements, SMS would in all likelihood have remained a marginal part-technology, and not the cultural practice it has become. The example of SMS interconnectivity is only one case where the dream of transparent and ubiquitous communication actually took a lot of technical and political work. Actors (and absent actors) that make networks fail to subsist can be technical, political, aesthetic and economic.

Contention over mobile standards has been the norm around the world (Agar, 2003). The apparently simple ideal of 'any to any' communication, which has become conventional wisdom in telecommunications, is nowhere near as straightforward as it seems. Infrastructures that are apparently ubiquitous always have to be achieved and maintained by localizers and articulators of many kinds: agreements on standards; material components; billing mechanisms; market research; promotion campaigns; subscriber enrolment, and so on. Advertising for mobile services presents a panoramic image of the nation and the family under the reassuring cover of connectivity. However, beyond zones of hyperconnectivity such as around Telstra Stadium, there are many dead zones, even in the city, and, beyond, huge expanses where mobiles have no connection.

In Our Own Image

Once the U2 show itself started, a significant number of people in the audience started using their mobile phones in a different way. They held them up, pointed them towards the stage, and began recording the show as still images or video. Most phone videos tend to be short single takes in low resolution, with the inevitable wobble from hand-held operation. Because of the limited storage available, the images are highly compressed, with prominent artefacts from the encryption process. The sound is invariably distorted, with the screams of nearby fans overriding the music. Videophones therefore produce a cinema of convenience, not a deep relationship with the moving image.

However, the differences from traditional video or film do not make these images unimportant. These non-human actors were full participants in sensations of the present, and part of the shared experience. The first audience for the images was most often the camera-movie-maker herself or himself, moments after the camera was switched to playback mode. The second audience was that person's friends, who watched the clip of what has just happened on the phone's small screen. The expense and complexity of sending clips elsewhere possibly precluded their further immediate electronic circulation, for now. There was little expectation that these files would be stored for a long period: unlike traditional photography. These observations of phone cameras at U2 seem to confirm Rivière's 2005 research that found phone cameras are part of a diversification of meanings of acts of photography. She argues that they increasingly capture the pleasure of the present moment, with less importance being placed on preserving something for the future or keeping a permanent record.

The pre-eminence of the present was also apparent in the lack of apparent concern from authorities about these recordings. The conditions of entry specified that cameras and recording equipment were not allowed into the stadium, presumably to protect the intellectual property rights of U2. However, the website specifically stated that patrons could bring in mobiles. Do mobile phones qualify as cameras? It seemed that neither the surveillance systems nor the security guards were policing this distinction: the chain of actors that may have regulated this behaviour did not extend far. The images, with their low quality, and their participation in the present, were certainly operating according to a different economy from traditional copyright.

In the weeks following the show, some of these phone and digital camera images circulated further from the stadium, as phone features such as Bluetooth, USB, and memory cards allowed files to be copied and sent elsewhere. The images were available most easily on the Internet, where photo-sharing sites such as Flickr, and video-sharing sites including YouTube made these files and media streams available around the world. Many personal blogs featured images and reports on the shows. Web visitors could access these artefacts by searching and browsing through tags, keywords and descriptions, or possibly following direct links sent to them by e-mail. Someone who was interested in the U2 show would get a very different impression watching these clips by comparison with the professionally edited live videos. They would see a fragmented composite of amateur images from different positions, evidence of a multiplicity of experiences.

However, in the same week as the U2 shows were taking place in Sydney, legislators in the federal parliament were putting in place their own programmes of action aimed at restricting copyright infringements. The *Sydney Morning Herald* ran a front-page story on legislation which would go before the Australian Senate in December. The story warned that the Copyright Amendment Bill proposed to make many of the fans' uses of camera phones illegal: even without knowing the law, offenders could have an on-the-spot fine, and even a criminal record (Murray, 2006a). While the government partially backed down on some details, such as allowing fans to make recordings of a performance, the revised legislation made it a crime to show recordings publicly, including playing them to friends or uploading to YouTube (Murray, 2006b). The question of whether these laws are enforceable has yet to be answered, when the network of actors for producing and distributing images has become so dense and extensive.

Becoming a Star

Towards the end of the show, the mobiles around the stadium made a final hurrah. Lead singer Bono incited the audience to make quite a different use of their mobile phones—to turn them on, hold them up, and 'set the stadium alight'. The house lights were turned down, and thousands of phones appeared around the stadium. U2 fans refer to this part of the show as the 'Milky Way'. It was a democratic light show in which every phone's backlight became a star. Through the song the majority of people attending held up a phone, almost as a religious icon.

Meanwhile the band played a melancholic anthem called 'one'. 'We're one, but we're not the same ... ' The meanings of this song are ambiguous. It was released in 1992, so some speculate that it relates to the reunification of Germany. Others claim that it is about someone dying from HIV/AIDS. The lyrics on their own could be read simply as referring to the existential problems of being separate individuals in a partnership. The song offers mainly atmosphere rather than meaning, which is stripped to the title: 'one'. This invocation to unity aspired to enrol the entire audience to its programme, one by one where possible, using all technologies available: pyrotechnics, multipurpose screens the size of buildings, lights and even a kite. The audience answered with their phones. This effort was in the tradition of the technologies that Latour calls panoramas, which present the most total image of the world by closing itself off from anything outside. The event answers

> a desire for wholeness and centrality. It is from those powerful stories that we get our metaphors for 'what binds us together', the passions we are supposed to share, the general outline of society's architecture, the master narrative with which we get a commonsensical idea that interactions occur in a 'wider' context: that there is an up and a down; that there is a local nested inside a global; and that there might be a Zeitgeist the spirit of which has yet to be devised. (Latour, 2005, p. 189)

Following the song, Bono moved into a monologue, reminding the audience that there are many things that you can do with a mobile phone—from ordering pizza to calling the Prime Minister: 'He works for *you*'. He then asked everyone with a phone to send an SMS message to support the 'make poverty history' campaign for policy change and charity to eliminate extreme poverty. The number miraculously appeared on the huge screen behind him. The phones descended from the Milky Way, and dimly lit thousands of faces. Each spectator turned their attention from the stage to the familiar image of their own phone, which accompanies them everywhere. And more, this programme demanded a meaningful task, of writing one's own name, and sending that in support of an important cause. At each show the system receives thousands of messages, handled by a chain of actors attributed to Intel. A few minutes later, some of our names begin to scroll across the large stadium screen.

While many enthusiastically wrote and sent their name to Bono, some people at the concert found this request unnerving. It demonstrated a capacity of technologies of mass individuation to capture so many names at such speed, even if those who submitted messages did so voluntarily. These 'oligoptica' (Latour, 2005, p. 181) do not comprise a total system. Rather, each individuated chain of actors maintains 'sturdy but extremely narrow views of the (connected) whole … as long as the connections hold' (p. 181). The following day, I received an SMS from Bono, asking me to visit the make poverty history website. U2 made use of a full repertoire of media to make connections with their audience. This canny translation from panorama to oligopticon is an indication of the domains of innovation opened up by technologies that operate at the massive scale of Telstra Stadium, and also make connections through the individuated level of mobile phones.

U2's incorporation of mobile phones into the finale of their performance is acknowledgement of changing audience expectations with the proliferation of individuated mobile technologies. The band has a reputation for reinventing itself. Using new media, often deliberately to excess, has been a big part of these practices of constant transformation (Johnson, 2004). On the other side, fans' tactical uses of mobile phones are changing their experiences with many 'mass' media forms by distributing attention and increasing the mix of modes of connection. Of course, these changes are happening not only at stadium shows, but with television programmes like *Big Brother*, alternative reality games, viral marketing, and so on. These diversified and individuated media bring not only different aesthetics and relationships between producers and audiences but also new revenue streams, ethical challenges and regulatory complications. As the fans streamed out of the stadium, the mobile phones were relatively quiet. Some were seeking a lift home, while others played back videos of the show. But I had seen enough to get a sense of the increasing complexity at play in events of mobile phone use.

References

Agar, J. (2003) *Constant Touch: a Global History of the Mobile Phone*, Icon Books, Cambridge.

Augé, M. (1995) *Non-places: Introduction to an Anthropology of Supermodernity*, Verso, London and New York.

Cunningham, M. (2005) 'U2//Vertigo//2005//Tour', Total production, Leander and Burbank US, Available at: http://www.totalproductionus.com/sitefiles/May05_U2.htm (accessed 25 Jan. 2007).

de Souza e Silva, A. (2006) 'Interfaces of hybrid spaces', in *The Cell Phone Reader: Essays in Social Transformation*, eds A. Kavoori & N. Arceneaux, Peter Lang, New York, pp. 19–43.

Donner, J. (2005) 'The social and economic implications of mobile telephony in Rwanda: an ownership/access typology', in *Thumb Culture: the Meaning of Mobile Phones for Society*, eds P. Glotz, S. Bertschi & C. Locke, Transcript, Bielefeld, pp. 37–52.

Gopinath, S. (2005) 'Ringtones, or the auditory logic of globalization', *First Monday*, no. 10, Available at: http://firstmonday.org/issues/issue10_12/gopinath/index.html (accessed 25 Jan. 2007).

Harper, R., Palen, L. & Taylor, A. (eds) (2005) *The Inside Text: Social, Cultural and Design Perspectives on SMS*, Springer, Dordrecht.

Horst, H. A. & Miller, D. (2006) *The Cell Phone: an Anthropology of Communication*, Berg, New York.

Johnson, F. (2004) 'U2, mythology, and mass-mediated survival', *Popular Music and Society*, vol. 27, no. 1, pp. 79–99.

Latour, B. (2005) *Reassembling the Social: an Introduction to Actor-Network-Theory*, Clarendon, Oxford.

Ling, R. (2004) *The Mobile Connection: the Cell Phone's Impact on Society*, Morgan Kaufmann, San Francisco.

Murray, L. (2006a) 'Hold those phones, rockers, soon your recordings will be a crime', *Sydney Morning Herald*, 14 Nov., p. 1.

Murray, L. (2006b) 'Rock, karaoke fans to play on after backdown by the copyright police', *Sydney Morning Herald*, 30 Nov., p. 5.

Patching, A. (1999) *Stadium: the Project Director's Diary*, Allen & Unwin, Sydney.

Rivière, C. (2005) 'Mobile phone cameras: a new form of "being together" in daily interpersonal communication', in *Mobile Communications: Re-negotiation of the Social Sphere*, eds R. S. Ling & P. E. Pedersen, Springer Elsevier/Morgan Kaufmann, London and Amsterdam, pp. 167–186.

Taggart, S. (2000) 'Oz: let the communications begin', *Wired News*, Available at: http://www.wired.com/news/technology/0,1282,38230,00.html (accessed 25 Jan. 2007).

Taylor, A. S. & Vincent, J. (2005) 'An SMS history', in *Mobile World: Past, Present and Future*, eds L. Hamill & A. Lasen, Springer, New York and London, pp. 75–91.

Vincent, J. (2005) 'Emotional attachment to mobile phones: an extraordinary relationship', in *Mobile World: Past, Present and Future*, eds L. Hamill & A. Lasen, Springer, New York and London, pp. 95–104.

Snapshots of *Almost* Contact: the Rise of Camera Phone Practices and a Case Study in Seoul, Korea

Larissa Hjorth

Snapshot: Introduction

As a burgeoning component of visual media cultures, camera phone practices are arguably both extending and creating emerging ways of seeing and representing. In media footage of late, camera phones have been heralded as providing everyday users with the possibility of self-expression and voice in the once one-way model of mass media. Building on discourses of analogue photographic practices and a so-called democratizing of photographic media, camera phones are affording users with the ability to document, re-present and perform the everyday. Moreover, camera phone practices also extend historical ideological constructions around bourgeoisie performances of intimacy and family that were initiated with the introduction of the Kodak camera. As Lisa Gye notes, camera phone advertising draws on the rhetoric of Kodak camera campaigns whereby consumers are compelled to document an occasion to make it 'worthy' of remembering (Gye, 2005).

In particular, the 'exchange' and gift-giving economy underpinning mobile phone practices (Taylor & Harper, 2002) is further enunciated by the camera phone's function of 'sharing' moments between intimates (and strangers) through various contextual frameworks and archives (from MMS, blogs, and virtual community sites to actual face-to-face digital storytelling). With the rise in convergent media through the interactive, pervasive, personal of mobile media, we are seeing everyday users creating their own digital storytelling techniques and diverse networks of distribution that further instil that *context* is one of the key factors in determining *content*.

With almost all mobile phones now coming with cameras, many users—not necessarily interested in photography per se—are becoming avid practitioners in the making, circulating and socializing of their own images. In the global information and communication technologies (ICTs) market, South Korea is being recognized as a key global player in creating innovative mobile technologies. So amongst the hype about the 'prosumer' (producer + consumer) and the democratic potential of mobile media, such as the camera phone, one wonders what the reality for users will be. Will, for example, female users turn around mass media's objectification of women? Or will users just perpetuate the same images, same stereotypes? Are camera phone practices creating new forms of storytelling and image making—or are these a reconstitution of earlier photographic practices that have their history in the analogue?

It is with these questions in mind that this paper will explore the burgeoning field of camera phone practices in one of the so-called global centres for mobile media innovation—South Korea. However, before exploring a sample group of Korean university students I will discuss the context of Korea as a global player in the new technologies industry. I will then entertain some of theories around new media and mobile communication informing the rise of camera phone practices. Finally I will focus on a case study of 34 Korean university students.

Getting Connected: Locating South Korea

In South Korea, net and mobile telephonic spaces are helping to 'progress' Korean forms of democracy (Kim, 2003a, p. 325). For Korean sociologists Shin Dong Kim (2003a, b) and Haejoang Cho (2004), the rise of a specific type of democracy in South Korea was afforded by, in part, new technologies such as mobile phones. In particular, in Seoul one can find two types of youth sociality predicated around two convergent technological spaces: firstly, that of the mobile phone (*hand phone*) and, secondly, that of the Internet through virtual communities (such as Cyworld's mini-hompys) and online multiplayer games and their attendant social spaces (that is, *PC bangs* meaning PC rooms). This usage of technological spaces is *not* about substituting the virtual with the actual but rather these technological spaces function to further connect individuals in the actual. The relevance of the technology is linked intrinsically to maintaining face-to-face social capital.

As a Digital Multimedia Broadband (DMB) and broadband 'centre', South Korea represents a prime example of innovative and emerging convergent mobile technologies. From September 2005, the streets of Seoul were coloured with posters advertising a DMB mobile phone content competition by TU mobile. Utilizing traditional Korean rainbow-like colours (that are also—coincidentally—similar to a TV test pattern), these ads were clearly converging Korean national identity with its prescient state as 'global DMB centre'. Here, the role of technology is bound up in the way in which Korea exports itself globally and consumes local products internally; the consuming of mobile technologies is tightly bound to explicit and implicit forms of nationalism. The hardware and software components are made in Korea, serviced by

Korean telecommunication companies, and with a population of 48 million and conservative estimates of 78 per cent penetration rates, the Korean mobile phone success story has taken global centre stage, drawing attention away from the once lauded Japanese DoCoMo *keitai* (mobile phone) IT revolution (Matsuda, 2005).

However, Korea's importance in manufacturing, branding and consuming mobile technologies provides an interesting case study of a country quickly accelerating into twenty-first-century modernity. It is incredible to consider that just 10 years ago the IMF bailed Korea out, and yet now, thanks to its leadership in technological innovation and production, Korea has secured itself as a source of cultural capital both globally and, specifically, the Asia-Pacific region. By scripting itself as a centre for digital, convergent technologies, Korea is rewriting its history through the global currency of ICT production. As Chua Beng Huat notes in his discussion of the trans-Asian flows of products and cultural capital in the region, the boom in 'consuming Korea' is part of a growing phenomenon be dubs 'communities of consumers' (2006). A revising of Benedict Anderson's (1991) notion of 'imagined community' to describe the associations and perceptions surrounding identifying with a nation-state, Chua's geo-political imaginary demonstrates that the region is constituted by identification and affiliations through commodities rather than through nation-states. This is particularly prevalent in the case of web 2.0 and the rise of social software.

In March 2006, Samsung released a 10 mega-pixel camera phone in Korea that would revolutionize the 'digital divide' surrounding quality, and thus content, between camera phones and stand-alone digital cameras. One of the dominant differences in the relationship between the two was linked to the depiction of, and association with, *official and unofficial occasions*. The former (camera phone) was always there, 'on hand' (both literally and metaphorically) to capture the trivialities of the everyday, unlike the stand-alone camera with high resolution that would be brought along purposely to events deemed 'special'. The second difference related to the *context* for sharing; the camera phone had 'sharing' built into its logic with quick functions such as MMS, Bluetooth or uploading to a blog almost instantly, whilst the digital camera would be taken to an often stationary computer and then uploaded. The launch of the 10 mega-pixel camera phone represented the *connection* of both these worlds. No longer would the camera phone images just be trivial and 'fun', they had the potential to be printed in high resolution, blurring the world between amateur and professional digital photography. Would this signal a change in the function and relationship to the camera phone? The implication of this convergent technology is, for everyday users, still far away, with the costs of such high technologies being out of the reach of the average user. For the moment, anyway.

As a burgeoning centre for innovative technologies and with a conspicuous usage of technologies in the everyday, South Korea's capital Seoul could be viewed as indicative of techno-nationalism. The projection of 'dynamic Korea' is one that has infused notions of technological innovation with the rise of power associated with Korea as a nation. With over 20,000 *PC bangs* gracing second levels of most commercial buildings, and with over one-third (18 million) of South Korea's population

(48 million) spending hours online in the virtual community of mini-hompy, one could be mistaken for believing that online identity and relationships were surpassing offline sociality. And whilst Koreans do, in general, have much trust in technological spaces such as the Internet as a site for reliably and democracy, the online is still no substitute for offline sociality.

As Kyongwon Yoon's ethnographic study of young people's use of mobile phones notes, the mobile phone helps to reinforce physical contact and exchange (Yoon, 2003). In Hjorth's and Heewon Kim's 2005 ethnographic study on youth using Cyworld's mini-hompy community, it was found that virtual connecting was always about the *need* and desires to be *connected* at various levels and *never* about *substituting* for face-to-face contact. In Florence Chee's persuasive ethnography on *PC bangs* and the politics of online multiplayer games, she argues that these spaces are *social spaces* that are viewed as 'third spaces' between home and work (Chee, 2005). For the youth of Korea, where most still live at home before getting married, these *third spaces* operate as spaces to connect with other people.

In these examples we see that the underlining logic of technological uptake is about the practice of sociotechnologies (Ito, 2005). With the technological innovations of mobile phone convergent media becoming increasingly more sophisticated, such devices as camera phones are no longer the 'poorer cousin' to the stand-alone digital camera. It is important to conceptualize mobile media as a genre within new media and thus contextualized through the notion of remediation.

In a period marked by convergent media and so-called interactive media, Bolter & Grusin (1999) utilize the notion of 'remediation' to make sense of the often non-causal relationships between new and old media. Just as Lev Manovich dismantled binaries between old (analogue) and new (digital) technologies in the early 1990s, Bolter and Grusin repurpose Marshall McLuhan's (1964) claim that the content of one (often newer) medium will also be that of another (often older) medium in order to transgress the old versus new debate. This notion of remediation is particularly prevalent when analysing the rise of mobile media techniques such as camera phone practices that feed into already existing image-identity making patterns that are regulated by the local.[1]

Parallels can be made between the rise of the webcam and 'reality' aesthetics associated with the hand-held camera in television and film, on the one hand, and the rise of the camera phone and sharing Internet communities such as Cyworld, MySpace and YouTube, on the other. As a convergent communicative medium premised on the logic of gift-giving, the various ways in which camera phone images can be 'stored', 'shared' and 'saved' is, as Ito and Okabe (2005) note, important to how the remediated sociotechnological images are read and contextualized. With the low resolution further giving 'authenticity' to the modes of realism ensured in the 'voice of the people' aesthetics, the camera phone signifies a glimpse into one's personal world. Here we see echoed the debates about new modes of stylized realism that can be witnessed in the rise of reality TV programmes, documentary/mockumentary genres and cinematic styles such as dogma 95.

As Finnish researcher Ilpo Koskinen notes, mobile multimedia is ordered by banality aesthetics in order to establish the sender as 'ordinary' and thus *reliable* (2006). For Koskinen, the implications of MMS are multiple and situated around 'the mundane as a problem' (2006, p. 50). This can be seen as a furthering of the politics of the everyday and personal as legitimated by Kodak snapshots. Certain archetypical images of homes and family occasions were an important part of presenting a family as normal both to themselves and others (Gye, 2005; Bourdieu, 1990). Moreover, the significance of mobile multimedia, unlike mobile telephony, is that it 're-territorializes' experiences and communication (Koskinen, 2006; Scifo, 2005). This emphasis on place underscores the motivations for documenting and sharing camera phone images through both online and offline means.

In the case of Korea, Dong-Hoo Lee has conducted ethnographic work into the ways in which camera phones both reinforce and yet provide contingencies for female empowerment and new ways of seeing/presenting (2005). As Lee notes, by 2004, mobile phone penetration rates where around 75 per cent, with 36.1 million people (Korea's total population is 48 million) having one or more handsets. Of these handsets 73 per cent of the hand phones sold in 2004 were equipped with built-in digital cameras; by the beginning of 2006 it was virtually impossible to buy a hand phone without the integrated camera.

As Lee notes in her previous study with Sohn (2004), the changing representational codes and accessibility for image-making practices and distribution are affording demographics that were previously excluded from the domain, most particularly women. In their study, Lee and Sohn found that women are more 'active in adopting new multi-media functions of the mobile phone' and that 'their willingness to adopt such functions is significantly stronger than men's' (2004, p. 243). For Lee, the history of photographic practice, and especially female representation in media, has been predominantly structured by what Lee defines as a 'patriarchal' society (revisiting Laura Mulvey's 'male gaze' argument). However, its recontextualization through the frame of the mobile phone (and the gendered nature of telephonic history and the gendering of private/public spheres) sees new emerging modes of self-presentation and identification that both, paradoxically, reinforce, and yet subvert, gendered imagery in mainstream media.

In the light of the changing, remediated nature of camera phone practices on genealogies of personal photographic discourses, such seminal research as Ito's and Okabe's in Japan and Lee's in Korea have sought to identify some of the key motivations and modes of storytelling by specific users. Continuing on from this empirical research, I will now turn to the micro-politics of the everyday user in a case study of 34 Korean university students conducted from September to December 2005.

Snapshots: an Ethnographic Case Study of Korean University Student Camera Phone Practices

During September to December 2005 I worked with 34 university students at Hallym University, asking them to analyse and comment on their own mobile phone practices.

The student group, coming from various disciplines such as English, Media and Communication, Sociology, IT, Film/TV, consisted of roughly half female and half male participants aged from 18 to 29 years old. Most were born in Chuncheon, apart from two Filipino exchange students. In a series of workshops in which students worked individually and collaboratively to compare and contrast, I asked students to reflect on their textual, aural and visual uses of the hand phone. Beginning with lectures on the ways in which theorists and practitioners have explored mobile media and ICTs, the seminar then moved onto three workshops in which students where asked to conduct three presentations. In the first presentation students worked individually to analyse their own camera phone images considering notions/motivations such as Ito and Okabe's three s's—sharing, storing and saving. The second presentation consisted of students working in pairs whereby each partner documented their own camera phone images for a week and then gave their data to their partner for further analysis. In the third and final presentation students documented their hand phone practices for a week and then gave the data to their partner for further analysis. This presentation also included students conducting interviews with partners after analysing the data to draw more personal meaning from the material.

This sample study of Korean university students is just that—a 'snapshot'. It is not meant to be representative of Korean young people in general. Rather, it is an attempt to sample actual users rather than just perpetuate media stereotypes about Korean techno-savvy young people that Yoon dubs 'neo-Confucian cyberkids' (2006). Moreover, since this study was 'controlled' in terms of being conducted at a university during a semester, one could argue that students regulated and censored some of the material. However, it should be noted that over the four months the students developed trust and friendships that led to much more open discussion of images and types of sharing than earlier in the semester; also, I had students working in collaboration and individually to collate and analyse the data. As they learnt forms of reflexivity, many students became skilled at constructively analysing fellow students' mobile data with great sensitivity.

Although this type of short-term ethnographic study is by no means unproblematic, it is one step in the direction of the much-needed quantitative longitudinal research in the area of mobile phone practices. In this way, the methodology of this study could be seen as a time-lapse snapshot. By working with the students for four months I was able to see how the mobile phone operated spatially and temporally in their everyday practices rather than a cursory glance by way of an interview or focus group study. Of course, I was mindful of the possibility of coercion and so I tried to give the students loose parameters so that the discussions and research could be contingent on their particular practices.

In the workshop, all but two students had camera phones. For over 80 per cent of the students their current camera phone was their first. For these individuals, the camera facility was not a consideration in the purchasing decision; many just bought the camera phone because it was difficult to buy a phone without a camera. The gender of the recipient did inflect the types of images that were taken, but only marginally. For many of these users, the three s's (Ito & Okabe, 2005) of camera phone images were predominantly via peer-to-peer hand phones and secondly mini-hompy sites (with sites such as flickr barely registering). This had to do with a majority of respondents noting their dissatisfaction with the low resolution of the camera phone and hence its being used only for 'trivial' everyday events and moments.

Whilst gender differences were fused with age differences, the most telling distinguishing factor was whether students were single or part of a couple. If they were part of a couple, the hand phone became the property of the female partner. She not only chose what objects adorned the outside of the phone, but, more importantly, what content was inside the phone. The phone was colonialized with signs of being 'engaged'. Male user phones became feminized with images of their girlfriend and cute objects adorning the outside. The phone thus becomes an incubator for a relationship, a co-present vehicle collecting memories as well as reminding the male user of his girlfriend (whilst signalling to others that he is 'taken'). The phone becomes a symbol not unlike the engagement ring; with the added function of reminding the male that wherever he is, his girlfriend is there (and he must call his girlfriend). In the case of one male student—who featured his girlfriend's eye as the screen saver, she is perpetually watching him.

This featuring of partners on the screen was common amongst the few 'coupled' respondents in the seminar. In turn, this type of coupled customization took many forms, from external customization of the hardware with matching cute characters hanging from both partners' phones (often bought by the female) to internalized customization (marking of the software) with such features as screen savers and ring tones. The screen saver image of the girlfriend functioned in a similar way as an engagement ring; the always visible hand phone on the table with the image of the girlfriend reminding the respondents—and people around him—that he was 'taken'. Here we see the hand phone customization as an extension of hand/arm ritualization—the image of the girlfriend is like an engagement ring, the hanging characters are like a charm bracelet full of mementos.

However, for all respondents, the most 'telling' form of personalization was the 'colourings'; that is, the music tune that replaces the dialling sound that callers hear when ringing someone. Whilst many did not download games and animations but instead used those that came with the phone, it was the colouring function that all respondents had customized as it was believed that this made the greatest impression on others about the user's tastes and values.

In the content of the camera images there was overt use of objects as analogies for special moments and gestures of friendship and sharing. Whilst there were examples of the typical 'group of friends' shot, many of the participants seemed to prefer taking pictures of surrounding objects to create more connotations. For example, the ritual of sharing food with a best friend; or the first two beers ordered on the night of celebration with friends; the sharing of coffee between two best friends. The female users tended towards more images of the 'cute'—especially in the form of cute animals or cute-style photos of friends. However, the images of male users in couples seemed dominated by the cute, often as a result of their girlfriend controlling the content of their phone.

Another key feature of the camera phone imagery was the role of media mimicry: many images consisted of 'paparazzi'-style photos, either of the user or their friends. Often, portraiture and self-portraiture mimicked the conventions of mass media such as paparazzi, but occasionally it slipped into metaphoric gestures on the sublime nature of the everyday, in which objects became the subject of paparazzi-style subjectification. A guilty glass. A lost shoe. A lonely can. The imagery also conformed to Koskinen's 'banality' principle in that often they were evocative of the DIY realism of the everyday. To a stranger they might be seen as meaningless, although enticing in the voyeuristic reality TV/cinema *vértité* tradition.

For most of the respondents, camera phone genres fell into the following categories: everyday, special occasions/places, friendship/family, self-portraiture, and favourites. Whilst the categories were often generic, the usage of the three s's was an important distinguishing factor in *how* and *why* respondents took their camera phone images. One of the most enduring aspects of camera phone usage and categorization was the way in which respondents spoke about it as 'personal', thereby not only rendering and contextualizing the images in terms of earlier epochs of personal photography but also in terms of a type of realism—or more appropriately reel-ism—that ensured their images were viewed by others as ordinary and thus reliable (Koskinen, 2006). One of the important factors that could be identified in the sample group was that camera phone practices were remediated, and thus contingent on Ito and Okabe's three s's in order to provide context for the content.

The key features of mobile multimodality—personal, ubiquitous, miniature, interactive, pervasive (location aware), and social—translate into gendered aesthetics of performativity. As can be noted by detailing some key case studies of the mobile media seminar, we can see that gender does configure in various forms—from the way in which photos are taken, to the subject matter and mode of sharing. Whilst almost all the students did not use MMS but preferred the more community-based network of mini-hompy to upload their images, many respondents were far from enamoured with their camera phones. The gendering of both genres and modes of performing was overt, most particularly with females using the camera phone to re-present fragments as opposed to male respondents who were more likely to 'document' themselves in social contexts that often inferred some type of hierarchy.

One of the key contextualizing factors is the context in which the images are shared and stored that differs between male and female and also single individuals and couples. The dominant mode for sharing, storing and saving was via the user's mini-hompy pages. Whilst half the respondents took, on average, about three images per week, this would occur mostly in order to *share* rather than for personal consumption. This is partly due to the fact that as a very social culture, for young Koreans the features of multimodal phones are primarily for communication and sharing. Whilst self-expression was, along with 'documentation', of special events/occasions—the two dominant modes of camera phone use—these were always about a negotiation of the social and 'remaining 'true' to one's identity on- and offline. The sharing of camera phone images is about re-territorializing place and the importance of social networks. They are co-present monuments to the importance of presence.

The mobile phone, in this case study, is an extension of social rituals that are predicated on actual contact. Much of the customization operates as an extension of the user's identity: pink, cute phones are used by young females to 'perform' conventional gender roles with a twist. The mobile phone screen saver operates as a charm bracelet to signify moments and memories with friends and family. For 'attached' users it can operate like an extension of the engagement ring, signalling to others that the user is taken whilst also reminding the user of the constant co-presence of their partner. If Marshall McLuhan's assertion that media become an extension of the body is correct then the mobile phone has become an extension of many body parts. It is an extension of the ear and mouth (telephony), the eyes (camera phone), whilst also becoming an extension of the hand (texting). It is the context of the hand that many mobile phone researchers have explored; Sadie Plant (2002) identifies the changing role of the thumb as the thumb takes precedence over the index figure. In a culture where one doesn't necessarily wear a wedding ring, the phone has indeed taken the role of the wedding finger for these couple users.

Conclusion

Camera phone practices are underscored by modes of realism. We feel like we are being allowed into someone's personal world—the ultimate fetishization of the personal. This pursuit of the real is not an unmediated real (as if such a notion exists) but rather a *reel* that is a homage to the analogue *real*. As Gye (2005) notes, current camera phone ads replicate the ads of the Kodak camera whereby people felt compelled to 'capture' the moment to render it worthy of 'official' collective memories. And with the recent launch of Samsung's 10 mega-pixel camera phone, the binary between the 'professional' stand-alone digital camera and the 'amateur' camera phone will dissolve and a new type of realism will prevail; one steeped in the rhetoric of the 'personal', 'confessional' and the 'everyday'.

Gye's discussion of camera phone imagery drawing on the familial conventions initiated by the Kodak camera is particularly apt in the case of the sample study of Korean university students and the types of single or coupled modes of photographic performativity. For participants who were single, their friends became part of the metaphoric family, often predominantly emphasizing one type of gender performativity (i.e. in the case of male participants it was 'masculinity', in the case of female participants it was 'femininity'). For coupled participants the phone was pivotal in signalling and collecting the moments of the 'new' family.

Here we see how the mobile phone is pivotal in adapting already existing cultural priorities and rituals.

The mobile phone becomes a vehicle for reinforcing familial roles—whether that be actual or symbolic. Thus we could do much to understand contemporary and future possibilities of camera phone practices by looking at previous, even analogue, photographic modes. We can see the same echoing issues about the role of historicity and temporality, the same feelings of both drudgery/obligation and the ephemeral. Whilst each 'new' medium creates debate about the role of technology it is important to remember that the content of the new medium is always that of a previous one (Bolter & Grusin, 1999; McLuhan, 1964). As these university students demonstrate, the camera phone is not transforming their relationships or identifications, but, as an everyday medium, it is allowing students to be more active in ritualizing their social relationships and ties.

In relation to the dominance of what Koskinen dubs 'banal but reliable' (2006), examples of camera phone imagery could be found in the work of the Korean students. Here we see that the emphasis is less on the actual content and genres and more about the sharing of the content. How the content is shared is also a defining contextual issue, with many using the virtual community of Cyworld mini-hompy to archive and disseminate. This can be understood in terms of a broader shift in web 2.0 towards the user's choice of *context* informing the *content* and *distribution* of their images in a social network. Whilst the mobile phone globally may be seen as an example of growing 'networked individualism' and a symbol of late capitalism, it is also very much inflected by the local and, in the case of the sample study of Korean university student users, informed by sociality. But like all practices of co-presence and intimacy, frustrations around time and representation will remain unabated and technology will often be the scapegoat for enduring patterns of youth cultures. The camera phone only serves to remind us of the history of photographic practice, and its attempt to capture the moment has always been obscured by longing desires. Welcome to the world of moments as sequences and memories as wallpaper between intimate strangers.

Note

[1] It is not difficult to feel a sense of déjà vu when hearing current debates around camera phone practices. Only a decade ago, heated discussion surrounded the relationship between analogue and digital photography; now it seems that a similar dialectic is occurring between camera phone imagery and the 'stand-alone' digital camera. In 'The paradoxes of digital photography', written before the introduction of camera phones, Manovich argues that the 'the logic of the digital image' is 'paradoxical', that is, it 'radically' breaks 'with older modes of visual representation whilst at the same time reinforcing these modes' (Manovich, 2003). This leads Manovich to assert that whilst film may disappear, cinema will far from vanish as it 'acquires a truly fetishistic status'. The digital world is overwhelmed with analogue metaphors, as can be witnessed in applications such as Photoshop, iMovie, and Final Cut.

References

Anderson, B. (1991) *Imagined Communities*, Verso, London.

Bolter, J. & Grusin, R. (1999) *Remediation: Understanding New Media*, MIT Press, Cambridge, MA.

Bourdieu, P. (1990) *Photography: a Middle-brow Art*, Polity Press, London.

Chee, F. (2005) 'Understanding Korean experiences of online game hype, identity, and the menace of the "Wang-tta"', *Paper presented at the 'Changing Views—Worlds in Play' DIGRA 2005 Conference*, Vancouver, Canada.

Cho, H. (2004) 'Youth, Internet, and alternative public space', Paper presented at the 'Urban Imaginaries' *Asia-Pacific Research Symposium*, Lingnan University, Hong Kong.

Chua, B.-H. (2006) 'East Asian pop culture: consumer communities and politics of the national', Paper presented at the *Cultural Space and the Public Sphere International Conference*, 15–16 Mar, Seoul, South Korea.

Gye, L. (2005) 'Picture this: the mobile camera phones and family photography' Paper presented at the 'Vital Signs' conference, ACMI Melbourne, 8 Sept.

Hjorth, L. & Heewon, K. (2005) 'Being there and being here: gendered customising of mobile 3G practices through a case study in Seoul', *Convergence*, vol. 11, pp. 49–55.

Ito, J. (2005) 'The creative commons: intellectual property and public broadcasting—opportunities for common sense and public good' Lecture for the Alfred Deakin innovation series, Melbourne Town Hall, 8 May.

Ito, M. & Okabe, D. (2005) 'Intimate visual co-presence', Paper presented at *UbiComp 2005*, Takanawa Prince Hotel, Tokyo, 11–14 Sept., Available at: http://www.itofisher.com/mito/ (accessed 10 Oct. 2006).

Kim, S. D. (2003a) 'The shaping of new politics in the era of mobile and cyber communication', in *Mobile Democracy*, ed. K. Nyíri, Passagen, Vienna.

Kim, S. D. (2003b) 'Korea: personal meanings', in *Perpetual Contact: Mobile Communication, Private Talk, Public Performance*, eds J. E. Katz & M. Aakhus, Cambridge University Press, Cambridge.

Koskinen, I. (2005) 'Ambient sound in mobile multimedia', in *Proceedings of Communications in the 21st Century: the Mobile Information Society: Seeing, Understanding, Learning in the Mobile Age*, Budapest, Hungary, 28–30 April, pp. 232–241.

Koskinen, I. (2006) 'Managing banality in mobile multimedia', in *The Social Construction and Usage of Communication Technologies: European and Asian Experiences*, eds R. Pertierra & I. Koskinen, Singapore University Press, Singapore, pp. 48–60.

Lee, D.-H. (2005) 'Women's creation of camera phone culture', *Fibreculture Journal*, no. 6, Available at: http://journal.fibreculture.org/issue6/ (accessed 20 May 2006).

Lee, D.-H. & Sohn, S.-H. (2004) 'Is there a gender difference in mobile phone usage?', in Paper presented at the *Mobile Communication and Social Change International Conference*, Seoul, Korea, 18–19 Oct.

Matsuda, M. (2005) 'Discourses of *Keitai* in Japan', in *Personal, Portable, Pedestrian: mobile phones in Japanese life*, eds M. Ito, D. Okabe & M. Matsuda, MIT Press, Cambridge, pp. 19–40.

Manovich, L. (2003) 'The paradoxes of digital photography', in *The Photography Reader*, ed. L. Wells, Routledge, London, pp. 240–249.

McLuhan, M. (1964) *Understanding Media: the Extensions of Man*, Routledge & Kegan Paul, London.

Plant, S. (2002) *On the Mobile: the Effects of Mobile Telephones on Social and Individual Life*, Motorola, Available at: http://www.motorola.com/mot/documents/0,1028,296,00.pdf (accessed 15 Jan. 2005).

Scifo, B. (2005) 'The domestication of the camera phone and MMS communications: the experience of young Italians', in *A Sense of Place: the Global and the Local in Mobile Communication*, ed. K. Nyíri, Passagen, Vienna.

Taylor, A. & Harper, R. (2002) 'Age-old practices in the "new World": a study of gift-giving between teenage mobile phone users', in *Changing Our World, Changing Ourselves, Proceedings of the SIGCHI Conference on Human Factors in Computing Systems*, Minneapolis, SIGCHI, pp. 439–446.

Yoon, K. (2003) 'Retraditionalizing the mobile: young people's sociality and mobile phone use in Seoul, South Korea', *European Journal of Cultural Studies*, vol. 6, no. 3, pp. 327–343.

Yoon, K. (2006) 'The making of neo-Confucian cyberkids: representations of young mobile phone users in South Korea', *New Media and Society*, vol. 8, no. 5, pp. 753–771.

'Hol' Awn Mek a Answer mi Cellular': Sex, Sexuality and the Cellular Phone in Urban Jamaica

Tanya Batson-Savage

Jamaicans have dialled into cellular technology at an amazing rate. Currently, there are 2.7 million cellular phone subscribers in Jamaica (ITU, 2005), a country with a population of 2.6 million people. The cellular phone made its entry into the Jamaican market towards the end of the twentieth century but it made its largest impact from 2000 onwards, with the entry of a second mobile services provider Digicel Jamaica (previously the market had been controlled by a telecommunications monopoly, namely Cable & Wireless Jamaica). With Digicel Jamaica's arrival the cellular market became highly competitive and the number of cellular phone users mushroomed. As result, subscriptions are split across three mobile networks: Digicel Jamaica enjoys the largest share of the market, followed by Cable & Wireless, and the third player in the Jamaican cellular phone market, Oceanic Digital. The Jamaican's passion for talk seemed to have taken everyone by surprise, including the service providers. In 2000 it was thought that 400,000 subscribers would be the island's saturation point, but Jamaica now has one of the highest cellular phone penetration rates in the world, with almost 102 subscribers per 100 inhabitants (ITU, 2005).

With this kind of penetration, it is not surprising that cellular phone technology has been remarkably incorporated into the people's lived experiences. Cellular companies dominate the media landscape with sponsorship and advertising and the cellular phone has replaced the cigarette lighter as the way to show respect in the dancehall. Phones have generally replaced watches and are often used as light sources in darkened theatres. Additionally, many people make their living from the cellular phone and its related accessories in both the formal and informal economy. In the formal economy they are usually employed by the cellular companies or repair services, while a myriad of other Jamaicans peddle phone cards and accessories and even offer repair services on the city's sidewalks. Heather Horst and Daniel Miller (2006) point to the multi-layered impact that the introduction of cellular technology has had on Jamaican society. Horst and Miller give a detailed exploration of the cellular phone in the hands of low-income Jamaicans in both urban and rural Jamaica:

> The phone links downward to word-of-mouth reportage, sideways to the dissemination of information through the media and, when it reaches its limits as a technology, upwards to another form of communication, transport, which then extends its ability to bring people and information together. (Horst & Miller, 2006, p. 137)

Horst and Miller focus on low-income Jamaicans because the thrust of their study is whether or not the cellular phone can impact on the island's development (that is, to help to grow its GDP). What becomes clear from Horst and Miller's work is that the cellular phone has been fully incorporated into the survival strategies of low-income Jamaicans.

This study departs from Horst and Miller's work through its focus on the sexual lives of Jamaicans and how this relates to the cellular phone. Though Horst and Miller make reference to Jamaican's sexual lives, they approach it with no detailed analysis. This study hopes to explore this issue further, and in particular how the cellular phone has been incorporated into the construction of sexuality and in sexual relationships. In doing so it explores the economics of sex and sexuality and how it is related to cellular phone usage. Additionally, informants in this paper are not limited to low-income Jamaicans.

Sexuality and the Cellular Phone: a Multi-track Reading

This paper explores the cellular phone's incorporation into the discourse of masculinity and femininity. Building on Simone de Beauvoir's argument that one is not born a woman but instead becomes one, and Luce Irigaray's definition of woman as a lack, Judith Butler argues that sexuality is not a passive construct of society but is instead an identity that one assumes (1999). As such, one is not identified as feminine or masculine until one performs the traits that are associated with the gender. In that sense, sexuality becomes a performance. This paper interrogates how the cellular phone has been incorporated into that performance of sexuality. This performance is relative to the phone's use as a status symbol as well as the economic imperatives that underlie cellular phone usage.

In pursuit of this, the paper employs a variation of Curwen Best's 'multi-track reading'. Best argues that much of the analysis of Caribbean culture, and Caribbean music in particular, has ignored the technology involved in the cultural production. As such, he argues that a multi-tracked reading involves taking this technology into consideration:

> A multi-tracked reading ... seeks to engage cultural phenomena by reading them backwards, if you will, interrogating the production process, de-scribing, unmixing the cultural product. It therefore concerns itself with analysing individual tracks of cultural data to come to terms with the technology which is built into the production process. (Best, 2004, pp. 6–7)

While Best uses this idea for reading culture created by technology, this paper explores technology's use in a lived culture: I employ the idea of multi-tracked to mean looking at various elements inscribed in, or involved in inscribing, elements of culture. To see how the cellular phone is used in the construction of sexuality I consider stand-up comedy, song lyrics, music videos, and interviews, treating each as a different track along which culture can be read.

The primary text informing this study comprises interviews, conducted in Kingston, Jamaica (that is, Kingston and the urbanized area of St. Andrew) over a three-month period in 2006. Informants of both sexes were randomly selected and ranged in age from the late teens to late 40s. Most of the interviews were conducted in the informant's place of work, school or home.

Song lyrics provide the secondary text for this paper. Several of the songs analysed in this study were recorded within the last five years, thus coinciding with the cellular phone boom. Using dancehall music as a lens through which to examine cellular phone culture and the performance of sexuality in urban Jamaica is particularly important if we understand that dancehall as more than the physical space embodied by the dances but rather a culture—a lived experience for which the music is a form of expression and a cultural manifestation. As Carolyn Cooper argues, DJ lyrics provide a 'distinctly urbanized folk ethos', as they remain free of the 'moral' and 'scribal' mores of the middle class (Cooper, 1993, p. 136). Though the continued penetration of the middle class into dancehall as the genre gains additional worldwide acceptance now makes it questionable how free of middle-class morality dancehall remains, it does provide a lens through which sex, sexuality, and their relation to power and politics in urban Jamaica can be examined.

Stand-up comedy, music videos, and newspaper articles (both opinion columns and news reports) provide supplementary input for the analysis here. While the interviews provided access to cellular phones as a part of lived culture, music videos, song lyrics and journalism allow for an understanding of its impact beyond the personalized experiences. As with the creation of music, these 'tracks' are interpreted as the multiple layers that are mixed down in everyday life to create a culture. Exploring them then allows us to understand the multi-layered role of the cellular phone in Jamaica, mixing and remixing the different 'tracks' to achieve a deeper understanding of the interaction between the cellular phone and sexuality.

Intimacy, Relationships, and Sexual Treaties

For the informants interviewed for this study the phones ranged from those that offered the basic tools of making and receiving phone calls as their main functions to those with digital cameras equipped for capturing still and video footage as well as Internet access. For some, the cellular phone's main function was as a medium to keep in touch with family and friends. For others it helped secure their financial stability, allowing them to become mobile offices, or at least keep in constant touch with their sources of income. Yet even informants who declared that their cellular phone's main purpose was work engaged in personal relationships via the cellular phone. As such, the cellular phone's capacity as a tool for communication, particularly in the private sphere, emerged as consistently important.

The first element to be explored here is the cellular phone as an access point, whether for people in a relationship or those pursuing one. It therefore functions as an important tool in negotiating sexual treaties. Several informants noted that the cellular phone had a pyrrhic role in relationships, simultaneously promoting and distilling intimacy. One informant, Ad-Ziko, suggested that the phone has impacted on the space in which relationships take place, shifting some to 'cell phone world'. This world is maintained through phone calls and text messages without commitment to physical contact. Ad-Ziko also argues that the cellular phone can 'dehydrate' intimacy as the quick intensity of communication via cellular phones can condense intimacy to its 'elemental essence'. Rather than injecting life into relationships, over-dependence on the cellular phone, it seems, can drain it of its vitality. So, the cellular phone also has the power to disconnect lovers even as they remain in communication. The 'dehydration' comes from the lack of physical contact, which the technology cannot simulate.

Additionally, it seems that the cellular phone is unsure of its place in romance, particularly because of issues of trust or independence. Some informants noted that they would question the motivation for giving a phone because it could indicate that the person wants to keep track of them, or would read the gift as an exchange for sexual favours. As such, it is only acceptable if it has been requested. For others the phone is too steeped in practical uses to be deemed romantic. Of course, some informants indicated that as the cellular phone is often an expensive gift it has its romantic merits. The cellular phone can also be deemed to influence romance through text messages, which can provide empowerment through distance. Rochelle explains that it was easier to engage in 'dirt text' with her boyfriend as the distance of text message allowed her to say things she was uncomfortable saying.

Furthermore, the cellular phone impacts on questions of faithfulness and also impacts on the construction of sexuality. According to Barry Chevannes' study of boyhood in urban Jamaica, the Jamaican male is socialized with sex as a pivotal part of his identity. Chevannes argues that sex is used to first establish and later maintain manhood (Chevannes, 2001, p. 199). Chevannes further argues that boys are socialized into a masculinity that is heterosexual, promotes toughness, being the family's provider and being able to 'win and survive at all costs' (Chevannes, 2001, p. 158). As such, the promotion of rampant heterosexuality as the masculine ideal has resulted where male heterosexuality is often maintained through multiple sexual liaisons. Some informants interviewed for my study suggest that the cellular phone has helped to make this a more efficient process by allowing men the opportunity to access multiple women with less effort. Miller and Horst point out that a unique feature of the Jamaican cellular phone landscape is that many people have a large network of numbers stored in their cellular phones. Describing the feature as the 'link-up' (a Jamaican term meaning either connection or re-union), Horst and Miller argue that the cellular phone functions as a link to various relationships for business, family, opportunities for schooling or work or just to get help. Interestingly, this feature is very important to the phone's implication in the construction of sex and sexuality. It is this network of numbers that allows for negotiating and or initiating sexual treaties. Men often employ the cellular phone to effectively engage in multiple sexual relationships while women use it to help to keep them financially solvent by staying in touch with men who will support them financially (Horst & Miller, 2006, pp. 81–101), something that I will explore below.

As one informant, Tanya, argues, the cellular phone can be viewed as a way that technology makes this easier and makes being 'a player' (that is a person adept at having multiple partners) 'more efficient'. This discourse has extended to the public domain again through comedy. This is highlighted by the comic duo Ity and Fancy Cat who have as a part of one of their routines how women engage in surreptitious conversations before their sexual partners by speaking in code but acting normally. As one informant, Sheldon, explains, the cellular phone teeters on the balance in a relationship:

> There are both advantages and disadvantages to having a cell phone in a relationship. Advantages—people can keep tab of each other. A disadvantage of having a cell phone in a relationship—people can keep tab a each other. Yeah, dat a my problem. For some people who have multiple partners and deh wid a woman an man an ting, phone cyan gi dem whe.

Of course, the cellular phone provides a means by which partners can both keep track of each other as well as investigate whether cheating is taking place by checking text messages. Several informants spoke of this trend and admitted to either using it or knowing people who used it. Some even admitted to having been caught by it. Horst and Miller point to how the cellular phone has gained a reputation as both a tool for duplicity and for ferreting out duplicity in sexual relationships (Horst & Miller, 2006, p. 98). This bears out Horst and Miller's observation that the 'subterfuge' mediating relationships can 'transform the phone from a device that connects into a device whose importance lies in its capacity to keep multiple strands separate' (Horst & Miller, 2006, p. 83). Interestingly, although Horst and Miller explain that this goes well beyond sexual relationship, many important instances of this, including the example they give, are still mediated around sexuality. Thus, the phone's ability to allow its user to effectively mediate among multiple relationships is particularly important to the construction of the masculine ideal.

Dialling into Dancehall

The cellular phone's role in negotiating sexual treaties and thereby functioning as a marker of sexual desirability can also be explored via dancehall music. In the song texts drawn upon here, women are the ground upon which masculinity is erected, allowing the man to display sexual prowess. Indeed, sexual prowess as a marker of masculinity is one of the hallmarks of dancehall music. Additionally, its presence in the music speaks to its wider acceptance in society. Chris Salewicz and Adrian Boot define dancehall as 'the marriage of digital beats and slackness: that moment and music in which lyrics about guns, women's body parts and men's sexual prowess come together' (Salewicz & Boot, 2001, p. 173). This definition is at best very limiting, yet Salewicz and Boot do accurately point to dancehall's concern with male sexual prowess. Indeed, sexual prowess for women and men becomes a marker of their desirability, and importantly their masculinity and femininity. The ability to be 'wicked inna bed'—that is not merely able to sexually satisfy but also to actually sexually dominate one's partner—is discussed in often seemingly violent and explicit terms in dancehall. Like the discourse on a desirable cellular phone, male DJs often need to show that they have great staying power and desirable features that make them the men that women turn to in order to gain sexual satisfaction.

The link between masculinity and the cellular phone is possibly most explicitly stated in Vybz Cartel's 'U No Have a Phone', which details how the DJ uses the various types of phones and phone plans to access sex. In the chorus, the DJ asks the question 'How yuh fi ave a gyal / an yuh no ave a phone?' Though seemingly innocuous, this rhetorical question cuts to the core of the definition of manhood. As explained above, a part of being a man is being able to maintain relationships with women, even if it is only a case of serial sexual liaisons. To not 'ave a gyal' and therefore not have proof of heterosexual prowess, is to risk being seen as not a man. So, the DJ explains that in contemporary Jamaica having a phone is essential to gaining access to women as it is often via the cellular phone that sexual treaties are negotiated. Having a cellular phone therefore becomes a symbol of manhood as it demonstrates proof of access to women and therefore proof of manhood. The cellular phone's role in arranging sexual liaisons is illustrated in Bounti Killer's 'Cellular Phone'.

'Cellular Phone' provides an interesting example of the intersection between the cellular phone, sex and desirable masculinity, and highlights how the cellular phone is used to enact multiple sexual treaties. In this song, Bounti Killer, one of dancehall's most prominent DJs, declares:

> Excuse pass mi cellular phone
> Mek mi call Antoinette and Simone
> Tell dem no move a muscle I'm comin home,
> To make a trip down in a dem love zone

Here the DJ uses the cellular phone to declare his impending arrival to Antoinette and Simone, on whose bodies he will unleash the full onslaught of his sexual prowess. Indeed, the line 'excuse pass mi cellular phone' suggests that he is in the middle of a grand performance—which he is, as the full strength of his masculinity is on display. Proof of his hyper-heterosexuality is evidenced in the DJ's calling not one but two women to slake his sexual thirst. Additionally, he reveals no trace of emotion as it is the women who possess a 'love zone'; he is going there merely to visit, as the use of the term 'trip' implies. He further declares:

> Gyal affi bawl and moan and groan
> Mi start drink di ole tiger bone
> We use brush, brush wi shoes and fling weh stone
> So tell di man dem leave that alone.

Tiger bone, brush (short for Chiney brush) and Stone[1] are generally employed to increase male stamina during sexual intercourse. The DJ therefore illustrates that he has no need for those and so he dismisses the 'brush' and throws away the stone. However, being a desirable man means being able to sexually satisfy a woman, and so some help is needed—he would rather rely on marijuana, tiger bone, and Guinness (the brand name synonymous with stout beers in Jamaica). His assertion that he can make women 'bawl', 'moan', and 'groan' echoes other kinds of violence. It therefore suggests that the DJ is wicked inside and outside of the bedroom, which further asserts his masculinity.

Interestingly, the frequency with which the cellular phone rings can signal manliness in that it marks heterosexual desirability and therefore masculinity. For instance, in the song 'Ring Mi Cellie' Lexxus chants his status as a star, and therefore a desirable man, as he has money and all the trappings that go with it: 'hear mi pon di radio, / see mi pon di tellie, / ring, ring, ring, ring, ring off mi cellie'. Desirable masculinity means having economic power, and symbols of this power are often worn as a part of the outward apparel—the cellular phone being notable among these. Indeed, the cellular phone as a marker of economic access will now be explored further.

Sex and Cellular Economics

In a study of masculinity in dancehall through the British Link-up Crew,[2] Donna Hope (2004) explores the connection between materialism and the construction of masculinity, highlighting how symbols of wealth including jewellery, expensive clothing and cars are used to confer status. Hope points out that having access and control of beautiful young women who are also draped over the man's arm as markers of wealth are used to confer status on men. I would argue from my research that this now includes the cellular phone, though there are some complexities to be worked through here.

Horst and Miller explain that the Jamaican is rapidly approaching the point where the lack of a cellular phone will signal 'destitution' rather than 'poverty'. They further argue that due to the proliferation of cellular phones in Jamaica, having a cellular phone no longer marks one as being in the 'vanguard of consumerism' (Horst & Miller, 2006, pp. 50–59). While this is quite true, all cellular phones are not equal and as such expensive phones have the power to separate the haves from the have-nots. As such, having an expensive cellular phone becomes a marker of economic power and can affect sexual desirability, as it is a marker of economic power and upward social mobility. So, not surprisingly, as much as fancy houses, jewellery, cars and beautiful (and usually half-naked) women have featured in dancehall videos, cellular phones are also often worn as visual accessories, as prominent as diamond studded necklaces. The phones worn are usually small and clipped to the waist or indicated only by the earpiece attachment, thus signalling the presence of a Bluetooth connection. This act of display of wealth becomes a part of the performance of gender as it is not just materiality that is on display but also power.

Hope's study also highlights that though dancehall grants power by allowing its players to access power though alternative and often informal or even illegal means, it in large part engages in the same patriarchal discourse as the rest of the society. As such, the status granted to men because they have access to money or women is no different, though it is seemingly more blatant in dancehall which thrives on display and exaggeration. So it is not surprising, then, that even among middle-class informants in my study it was clear that the type of cellular phone was used as a marker of status. Clearly the use of cellular phones as markers of economic power goes well beyond song lyrics and music videos. Several informants point to this element, whether through their own experiences or through things they have observed. Leighton explains the difference between how he was perceived with his old phone which had only basic features and was once battered and had a cracked screen. Now his phone earns him respect from men and women:

> [S]ince I've been carrying this one, whenever I take it out, whether it's a male or female the attitude is different. Where there's a male they'll say 'Wha you have big phone man.' If it's a girl, almost instantly like, it's a pull especially if she falls below 23. Automatically, it's a pull. It's like you're rolling with the big times, when you have a phone like that.

The idea that the cellular phone acts as a 'pull' to women by signalling his access to economic power, that is 'rolling with the big times', highlights the cellular phone's symbolic power. For 26-year-old O'neil, a working-class man, the cellular phone either brings a man into contact with ridicule or affords him access to willing women:

> They say like, based on the type of phone that you have, you're like cheap! That phone cannot compare to fi dem because theirs might be can record, take pictures and stuff like that and do a lot of things weh basically your own cannot do ... The most expensive phone that everybody is going crazy over and these blue berry and these k700i and stuff like that these phones weh can do a lot of things like record, videoing and stuff like that, yeah once [women] know that you got a camera on it, is like they go crazy. [He laughs] I don't know why.

The perception that markers of economic power bring access to women does not appear to be a false delusion that men have, as some women admit to the value of the cellular phone as a marker of economic power. As 21-year-old Krista explains:

> [I]f it's a man and him don't have a nice phone you goin presume say him don't have no money, especially if him come in a club and try to pick you up, with one of them like old nokia or the siemens dem. You going to presume this man have no money,

> can't spend on anything, him not hype and him not with it. My boyfriend now says he's gonna buy a new phone.

The phone represents economic access and proof that its owner is 'hype' that is, interested in fun and the modern.

Due to gender dynamics, however, an expensive cellular phone being worn by a woman reads differently. It can be an indicator that she is financially independent, but it can also indicate that she has access to economic power via her man. As such, the cellular phone can mean she is either accessible or can only be accessed by men with the economic power to support her. Calcion, a 21-year-old student, notes that men whom she ignores on the street or in class assume she does this because they cannot buy her an expensive phone. The song 'Regular', sung by Miss Thing, adds additional insight into this perception and highlights the relation between sex and economics and the cellular phone. The song uses the cellular phone to mark a woman who is not 'regular', meaning that she is extraordinary and extremely desirable. The song begins:

> From you know seh you hot from yuh bawn
> Ole up yuh cellular an tun it aawn
> From yuh know you hyper dan Dawn
> Tun it aawn, tun it aawn.
> Yuh no common an big tings a gwaan
> Tun it aawn, tun it aawn.

With the first two lines the DJ draws the line between the cellular phone as a marker of economic access and sexual desirability, suggesting that intrinsic hotness ('hot from yuh bawn') comes with economic access. This access is not through the woman's own entrepreneurial activity but rather through her ability to get and retain a man of economic worth. This is even more clearly enunciated in the song's chorus, which highlights the connection between a woman's desirability, a cellular phone and having a man with economic access, as the DJ outlines: '[Yuh] A get man cause you no regular / from you have credit show me yuh cellular / comfortable cause yuh bwoifren a hustler / gyal bruk out an wine'. These lines highlight that the woman's comfort level can be determined by her boyfriend's entrepreneurial endeavours, be they legal or illegal, as hustle can refer to either. The song therefore reaffirms the man's place as the breadwinner while legitimizing the woman's use of sex to maintain her lifestyle. This ought not be confused with the promotion of prostitution, however, which is decried. Furthermore, the oft repeated 'tun it aawn' also echoes the woman's ability to turn on the man with the same ease with which she can turn on her phone. This ability to easily 'tun [him] aawn' is the true marker of her femininity. Chevannes highlights the fact that the image of the man as provider is intrinsic to the creation of the Jamaican male identity. He argues that from boyhood males learn that in a relationship they are expected to bear the brunt of the financial responsibility. He further explains that economic power comes right back to sex and sexuality:

> The ideology governing the gender relations prevailing in Joetown seems very transactional. Men need sex, which women have; women need money, which men must find to satisfy their demand. No money, no sex. In this view women need three

> things: good money, good treatment and good sex, in that order, with sex coming last. For the man, not so. Sex comes first. (Chevannes, 2001, pp. 198–199)

Here, Chevannes outlines the crux upon which sex and sexual economics operate. Just as men use the phone to access women, women use the phone to access men from whom they can get money, or at least credit for their cellular phone.

'Regular' also illustrates the competitive lines by which femininity is drawn, particularly in such sexual economics. The unnamed 'you' is touted as 'hyper dan Dawn' and so the song pits women against each other clearly demarcating them into women who have achieved their sexual ideal by being attractive and securing a man of their own and those who have to either share a man or sustain their lifestyle, or worse yet, turn to prostitution. In this dynamic, women sometimes engage in a battle over men who have become a part of the scarce benefits and spoils (Tafari-Ama, 2006). Tafari-Ama explains that a woman's power can come from the man she secures as her own, whether because of the economic power he brings or because of his own status in society. The phenomenon of the 'matey wars' has proceeded from this and has become a much-discussed element of popular culture. The matey, or the other woman, is a significant part of the Jamaican sexual matrix. From this comes the practice that it is manly to have multiple women.

Tafari-Ama points out that there is an economic impetus that allows some women to accept this as a part of life, and equally explains why some women are willing to go into physical battle with other women to secure possession of their man. She notes that the spectacle of the matey war (a physical or verbal battle over rights to a man) has become an accepted way of life and often provides much entertainment for passers-by who have even placed bets on the outcome. The matey war has also found a home in various song lyrics, and traditionally the matey has been treated with the scorn and disregard of the fallen woman.

However, as Carolyn Cooper has argued, particularly in reference to the work of Lady Saw and even Shabba Ranks, the dancehall can prove a sexually liberating space for women, often freeing them of the moral shackles to which the middleclass attempts to bind them (Cooper, 2004, pp. 73–123). As such, the patriarchal underpinnings of the definition of the feminine in dancehall are constantly in flux. The matey who is successful at stealing another woman's man is often praised for her skills. Indeed, dancehall musician Tanya Stephens has provided a radical redefinition of the matey as a woman secure in her sense of self and independence, so she can pick, choose, and even ultimately reject her man if he proves to be less than worthy either financially or sexually (Batson-Savage, 2005, p. 9)

The cellular phone has also become implicated in this phenomenon both at the level of music and in practice. Stephens' album *Rebelution* provides an excellent example of how the cellular phone can work in a matey war. In the interlude dubbed 'The Message', an irate girlfriend calls the DJ and threatens first her and the wayward boyfriend with extreme violence because of the public humiliation they have afforded her. 'The Message' was clearly left on a cellular phone because it uses the pre-recorded instructions used by Digicel for its messaging service. 'The Message' highlights the fact that the cellular phone has also affected the space in which these 'matey wars' take place, now becoming a space where the competing femininities of wife and matey collide.[3]

The cellular phone bill is another important element of cellular economics and the question of the cell bill has dominated much of the public discourse about the cellular phone surfacing in song lyrics, comedy routines and even newspaper articles. In one of his stand-up routines, popular comedian Owen 'Blacka' Ellis states that 'call me back' had become the three little words that replaced 'I love you'. He explained that when a woman calls a man, rather than quickly stating 'I love you' she says 'call me back', then hangs up. The hilarity with which the joke is often greeted suggests that the audience quite identifies with the scenario the comedian describes. Ellis' joke also points to the cutting edge between romance and finance. The phone bill ties in with such a definition of masculine and feminine roles, as it is often expected that the man should foot the bill, because he is the man, or because he wants to sleep with the woman. Or simply because she is a woman and therefore it is a part of his duty to take care of her. The length of time for which a man's credit lasts becomes a marker of his financial stamina, which he needs to assert his masculinity.

However, this idea is not all pervasive. Many men admit that they do not condone the behaviour, but that they often find that some women expect it of them. Some men explain it away by noting that while it is generally okay for them to supply a woman with credit, it would be nice if once in a while she returns the favour. Of course, for some women there is no question that they ought to pay their own bill, though they find it unmanly for a man to ask them for credit or send a 'please call me' text. As Rochelle, a 32-year-old informant who is adamant about paying her own cellular phone bill, puts it: 'if he cannot finance a phone call he definitely cannot finance a relationship'.

The Digicel feature popularly dubbed 'Please Call Me' through which a person with J$3 or less credit on their phone can send a text to another Digicel number asking them to return a phone call is one of the unique features of the Jamaican cellular phone market and, as Horst and Miller illustrate, it dovetails into how ordinary Jamaicans beg, trade, or give cellular phone credit. Since the feature's arrival in 2004, Digicel has since introduced the 'credit you' feature through which one can send credit to another person's phone, starting from a denomination of J$25. The 'Please Call Me' feature also has implications for the performance of sexuality, and reactions to it have implications for changing ideas about female economic independence. 'Please Call Me's' immediate popularity propelled it into public discussion in the media. Interviewed individuals in one article complain about women's abuse of the feature: 'It seems as if these girls nowadays never have any credit. They text you at all hours in the morning with these silly "Please call Me" signs, and then after you call them, they beg you to buy them credit'(Mills, 2004, p. 3). However, while many complain about abuse of the Please Call Me system, some do admit to it as a way of life. Others see the value of the feature in an emergency but will not tolerate its being used frivolously.

Interestingly, the system seems to be better tolerated when it is used either among friends or family. Some men do not complain about being asked to supply women with credit as they think that it is a part of their role as men to help take care of women. Hence DJ Vybz Cartel, in a song called 'Please Call Me', uses the feature to underline that he is desirable because of his long-lasting love making and long-lasting credit.

Conclusion: Aiding and Thwarting Intimacy, and Defining Selfhood

Journalist Mel Cooke remarks that the cellular phone helps us to deal with modernity, keeping us in touch in 'a more impersonal, hustle and bustle society' (Cooke, 2006, p. A6). The technology has the ability simultaneously to promote and thwart intimacy and though it is often featured in romantic liaisons is often not seen as romantic. Moreover, the cellular phone has effectively dialled into the discourse on sex and sexuality in urban Jamaica, touching on questions of selfhood and identity. Its arrival has not changed the way in which gender is constructed, but it has been integrated into its construction. In many ways the cellular phone has been used to reaffirm the patriarchal thrust of gender constructions, and has become a part of the physical symbolic construction of gender playing into the use of materialism to achieve selfhood. The socio-economic imperatives that guide cellular phone use also impact on gender construction, and the performance of gender traits such as expensive phones and long-lasting credit is used to represent the masculine ideal.

As I have tried to show, cellular phones have even impacted upon the spaces in which relationships are mapped by becoming a site for relationships and gender wars. The cellular phone's symbolic power comes from its multiple roles as a point of access, as well as the mark of sexual desirability. Although it has implications for femininity, its association with economic and sexual power allows it greater resonance with the construction of masculinity. Intriguingly, when asked whether they would date their cellular phone if it were a man or a woman, many informants either had or were looking for cellular phones similar to their ideal mate. This suggests that not only has the cellular phone impacted on the construction of sexuality but also our phones resonate with the men and/or women we would choose to date.

Notes

[1] Though all three are used to prolong a man's sexual performance, tiger bone, a liquid which is drunk, is perceived as a tonic and therefore more natural. Stone is a solid that is rubbed onto the penis and chiney brush is sprayed on. Both deaden sensation in the penis.

[2] The British Link-up Crew is a popular group of promoters who stage parties in the United States, Jamaica, and Britain.

[3] The 'wifey' and 'matey' can be viewed through the spectrum of the virgin–whore dynamic though such a binary opposition would distort the differences and similarities that exist between them. The wife (used interchangeably with 'wifey') is not necessarily married to the man but

is viewed as his steady partner. Furthermore, it is more faithfulness and good maternal instincts that are promoted in the 'wife', rather than chastity.

References

Batson-Savage, T. (2005) 'Gangsta poetics: femininity in Tanya Stephens', *Gangsta Blues*', *Jamaica Journal*, vol. 29, no. 1–2, pp. 6–11.

Best, C. (2004) *Culture @ the Cutting Edge: Tracking Caribbean Popular Music*, University of the West Indies Press, Jamaica, Barbados and Trinidad & Tobago.

Butler, J. (1999) *Gender Trouble: Feminism and the Subversion of Identity*, Routledge, New York and London.

Chevannes, B. (2001) *Learning to be a Man: Culture, Socialization and Gender Identity in Five Caribbean Communities*, University of the West Indies Press, Barbados, Jamaica and Trinidad & Tobago.

Cooke, M. (2006) 'Those flipping cellies', *The Gleaner*, 5 Jan., p. A6.

Cooper, C. (1993) *Noises in the Blood: Gender, Orality and the 'Vulgar' Body of Jamaican Popular Culture*, Macmillan Caribbean, London.

Cooper, C. (2004) *Sound Clash: Jamaican Dancehall Culture at Large*, Palgrave Macmillan, New York.

Hope, D. (2004) 'The British Link-up Crew: consumption masquerading as masculinity in the dancehall', *Interventions*, vol. 6, no. 1, pp. 102–117.

Horst, H. & Miller, D. (2006) *The Cell Phone: an Anthropology of Communication*, Berg, Oxford and New York.

International Telecommunications Union (ITU) (2005) 'Mobile cellular subscribers per 100 people', Available at: http://www.itu.int/ITU-D/icteye/Indicators/Indicators.aspx#

Mills, C. (2004) 'Please Call Me', *The Gleaner*, 27 Jun., p. 3.

Salewicz, C. & Boot, A. (2001) *Reggae Explosion: the Story of Jamaican Music*, Ian Randle, Kingston.

Tafari-Ama, I. (2006) *Blood, Bullets and Bodies: Sexual Politics below Jamaica's Poverty Line*, Multi Media Communications, USA.

Discography

Vybz Cartel (2006) 'U No Have a Phone', *JMT*, Greensleeves.

Bounti Killer (1996) 'Cellular Phone', *No Argument*, Greensleeves.

Tanya Stephens (2006) *Rebelution*, VP Records

Miss Thing (2004) 'Regular', *Miss Jamaica*, Sequence.

Mr. Lex (2000) 'Ring Mi Cellie', AKA Mr. Lex.

Overseas Filipino Workers and Text Messaging: Reinventing Transnational Mothering

Cecilia Uy-Tioco

Overseas Filipino Workers (OFWs) have been hailed by the Philippine government as 'new heroes'. In 2005, over US$10 billion were sent home in remittances, contributing to 10 per cent of the gross national product (Banko Sentral ng Pilipinas, n.d.). According to the Philippine Overseas Employment Agency (POEA), 933,588 Filipinos left the country for employment overseas in 2004 (POEA, 2004). This number represents a 7.6 per cent increase from the previous year yet is only a count of those who have sought overseas employment legitimately. Filipinos who seek employment illegally are not accounted for and their numbers are also on the rise. Of the almost a million documented workers, 75 per cent are women (POEA, 2004), many of whom leave children behind to be cared for by husbands and the extended family. Much of the labour provided by Filipino women workers is in the care industry. They enter the global workforce as nurses, domestic helpers, and nightclub entertainers, filling in the gaps created by the privileged women of the West. Many Filipino migrant mothers use cell phone technology to maintain relationships with their children: the technology impacts relationships and raises troubling implications. Cell phone technology has empowered these women, creating new ways to 'mother' their children across time and space. While allowing for deep-seated ideological beliefs of traditional mothering to continue, these mothers are also simultaneously reinventing transnational motherhood. Furthermore, this use of cell phone technology can be read as both a form of resistance against and repression from the political economic reality that has led these women to leave their families in the first place.

This paper is based primarily on observations and informal, open-ended interviews I conducted with Filipina overseas workers living in the east coast of the United States. These conversations were carried-out in Taglish (a hybrid of Tagalog and English) and translated into English. The names of the interviewees have been changed to protect their privacy. Because my focus is on cell phone use of migrant Filipino women in the Unites States, I rely on existing scholarly work on children of migrant mothers and transnational families, particularly that of Rhacel Salazar Parreñas, to provide insights regarding the children who have been left behind. In addition, narrative accounts in newspaper and magazine articles have been incredibly useful.

Raul Pertierra is the most notable scholar on mobile phones in the Philippines. His co-authored book *Txt-ing Selves: Cellphones and Philippine Modernity* (2002) provides an excellent overview of cell phones in the Philippines and illustrates how much they have become part of daily life—as well as discussing their role in ousting a Philippine president in 2001. In 'Mobile Phones, Identity and Discursive Intimacy' (2005), Pertierra further explores the effects of cell phones on identity, particularly sexuality and intimacy, as well as the cell phone's political possibilities. Text messaging, for Pertierra *et al.*, combines the intimacy of speech with the reflectiveness of writing (2002, p. 140). Cell phones are used to extend existing social relationships but also to expand social networks and create new forms of subjectivities (Pertierra, 2005, p. 28). Like Pertierra, I explore how cell phones and text messaging maintain and enhance existing relationships, especially intimate ones of mother and child. This paper, however, further examines questions of power relations and structures that are reproduced, rejected, or reinforced by cell phones and text messaging.

Bella Ellwood-Clayton's research on cell phones and dating in Kalibo, Aklan, in the central Philippines focuses on the effects of text messaging on traditional courtship practices and its 'potential to subvert traditional gender ideologies in the domain of young love' (2005, p. 226). In her research, Ellwood-Clayton proposes that text messaging is 'utilized as both a romantic tool and a vindictive weapon' (2005, p. 196). Through texting, women are no longer passive in the courtship ritual, but rather are able to explore new roles. Like Ellwood-Clayton, I recognize the liberatory potential of cell phones for women. However, I also assert that as the technology empowers, it reinforces patriarchy and the North–South divide.

While Pertierra and Ellwood-Clayton focus their research on mobile phone use in Philippine society, this paper examines the uses of cell phones and text messaging and their implications for a fragment of the Philippine diaspora—Filipino migrant mothers in the United States. Both Pertierra and Ellwood-Clayton provide invaluable information and analysis on the impact and ubiquity of cell phones in Philippine society. By focusing on Filipinos *outside* of the country, this paper complements their findings and magnifies the impact of cell phones on Filipinos everywhere.

Migrant Labour and the Filipino Family

It is ironic that while care is the largest export product of the Philippines (Parreñas, 2005, p. 24), caring for one's own family is sacrificed. Migration of a parent abroad is a way to maintain households economically; at the same time, it breaks up the family physically and geographically. While the country benefits from the income brought in by OFWs and hails them as national heroes, transnational families—especially when it is the mother who is away—are still seen as abnormal because they question what traditional families are expected to be like. The Filipino family is believed to be the foundation of the nation, and by 'family', emphasis is put on the nuclear family of two heterosexual, married parents and children. The continuity of the family is vital in Philippine society, placing high value on both marriage and children (Chant & McIlwaine, 1995, p. 16). Emotional nurturing, value-formation, and care for physical well-being are responsibilities attributed to mothering. Traditional mothering involves being physically present to feed, clothe, and provide emotional support to children. While migrant mothers are able to provide their children and families with financial wealth and economic stability, the price for women has been separation and alienation for children and a sense of loss and the guilt of not being a 'good' mother.

Although economics and the desire to give their children 'better lives' are what push Filipina mothers to seek employment abroad, alienation from the family is the greatest price they have to pay. Studies such as that of Parreñas have assessed the consequences that migrant mothering has on the Filipino family. Children often feel abandoned and while they enjoy the material wealth their mothers provide, it often does not compensate for the distance and the alienation they feel. When the families are reunited, relationships are often strained and difficult to rebuild. The children who have been left behind have either grown to teenagers or adults who are used to living without their mothers. Oftentimes children resent their mothers for their absence and mothers resent their children for not being grateful for the sacrifices they made to provide for them.

Children of migrant mothers especially feel the emotional strain forced upon transnational families. According to Parreñas, children in transnational families 'hunger for emotional bonds with absentee parents and wish for the intimacies of everyday interactions' (2003, p. 133). Despite the tremendous support that many children receive from their fathers or the extended family (many live with grandparents or aunts), the longing for a physically present mother remains strong. Junelyn Gonzaga, a child of migrant parents, points out that while she understands the sacrifices her parents made for her, they must also recognize the sacrifices that children go through (Parreñas, 2003, p. 141).

In her study, Parreñas asked a focus group of men and women who had relatives or knew of someone who was working abroad what they thought were the effects on the family of overseas migration. The responses were: families are neglected; children are abandoned; the attitudes of the children change, values disappear, and they take on vices; and men take mistresses (Parreñas, 2005, pp. 42–43). Unsurprisingly, there is a general belief that if a parent needs to migrate for employment, it should be the father since he is expected to be the breadwinner of the family. Thus, if a mother leaves for work abroad, leaving the care of her children to their father, it is believed that the children will not receive adequate care and nurturing. When children of migrant mothers misbehave or engage in typical schoolyard pranks, blame is often placed on the fact that their mothers are away. According to Parreñas, 'a central paradox of the maintenance of transnational families is the fact that the achievement of financial security for the sake of the children goes hand-in-hand with an increase in emotional insecurity' (Parreñas, 2003, p. 149). The economic benefit of OFWs comes at seemingly high costs to the traditional family unit.

In their pioneering work in 1997, Pierrette Hondagneu-Sotelo and Ernestine Avila defined transnational motherhood as 'circuits of affection, caring, and financial support that transcend national borders'. In studying Latina domestic workers in Los Angeles, Hondagneu-Sotelo and Avila examined the 'forging of new arrangements and meanings of motherhood' (1997, p. 549). The authors discovered that mothering ties are maintained through sending remittances home, providing for the financial needs of children. Coupled with this are the regular writing of letters and phone calls. Although letters and phone calls are not constant and are often few and far between, they regularly contain admonishments to take vitamins, to study well, and to mind their manners. The rise in cell phone technology reinvents this notion of transnational mothering, allowing for real-time and constant communication between mothers and children separated by time and space.

Cell Phones in the Philippines

Cell phones were introduced in the Philippines in the mid-1990s, almost simultaneously with the Internet. Since then cell phones have become important for everyone. No longer simply a gadget for the rich, but also a common appliance for the middle class and even the poor, cell phone adoption has crossed the boundaries of age, gender, and class. In 2005, the two main mobile phone companies, Globe Telecom and Smart Communications, were serving a total of 32.8 million cell phone subscribers in a country of 89 million people—36 per cent of the population. This number is significant because 40 per cent of the country lives below the poverty line on less than a US$1 a day (CIA, 2006).

Perhaps one of the biggest surprises in cell phone use is the adoption of Short Message Service (SMS) or text messaging as the primary method of communicating (for a history of text messaging see Goggin, 2006). Called the 'text capital of the world', the country's residents send and receive more cell phone text messages than citizens of any other nation (Diamond, 2002). In 2004, 200 million texts a day originated from the Philippines (quoted in Pertierra, 2005, p. 27). Anyone who encounters a Filipino and his or her cell phone will be amazed at the speed with which he or she can text. Because this add-on service has been around since the late 1990s, Filipinos have become adept at keying in messages using the telephone keys, some using both hands, some with just one hand and the thumb doing all the punching:

> Sending messages by phone is an irritating skill to master, largely because 26 letters, plus punctuation, have to be created with only 10 buttons. Typing the letter C, for example, requires pressing the No. 2 button three times; an E is the No. 3 button pressed twice; and so on. (Arnold, 2000, p. C1)

Although learning this skill is tricky, once adept the message is sent immediately to the recipient, who is equally as skilled in the process, allowing immediate communication. According to Levinson, the act of 'writing has always been more quiet than speaking ... and more precise' (2004, p. 106). So not only can text messages be sent at anytime, the sender is assured that he or she has got the message across. This is especially helpful when a message needs to be relayed but the sender is not sure whether the receiver is available to take calls. Instead of risking a voice call that can interrupt something, a text message is more silent and can be attended to at a later time.

The popularity of text messaging in the Philippines has been attributed primarily to its affordability. When GSM cell phones first entered the market in 1999, text messaging was a free, add-on service. As its popularity grew, a cap on free texts (dependent on the cell phone plan one subscribed to) was placed and additional text messages were charged to the sender. Users are not charged for text messages or voice calls that they receive. Despite this, the cost of text messaging remains considerably lower than that of voice calls. Besides affordability, there are other cultural and social reasons why text messaging has taken root in Philippines society. In *Television: Technology and Cultural Form*, Raymond Williams challenges us to look at technology not as self-acting forces that create new ways of life or provide material for new ways of life, but rather as created and developed for certain purposes and practices in mind, and that these are known social needs (1974, p. 8). For Williams, pre-existing conditions and practices in a culture allow for technologies to develop and be adopted. Lister *et al.* point out that for Williams, 'new technologies take forward existing practices that particular social groups see as important or necessary' (2003, p. 81). According to Pertierra *et al.*, 'the Philippines is primarily an oral society and speech is often used to maintain and reproduce traditional hierarchies. It is usually easer to challenge the restrictions of speech through writing' (2005, p. 140). Through text messaging, hierarchal barriers of class, age, and gender are broken down or, at the very least, challenged.

While the marketing and advertising campaigns of Philippine cell phone companies cannot be overlooked, there are certain pre-existing Filipino cultural practices that have paved the way for cell phone technology and text messaging to flourish. In previous research I suggested that these cultural practices are: the importance of family, the concept of *hiya* (loosely translated as shyness or embarrassment), and the need to be in the know (for a more detailed discussion on this see Uy-Tioco, 2004). The family—nuclear, extended, and transnational—is the cornerstone of Philippine culture, and cell phones have made keeping in touch easier, allowing for family bonds to be maintained or strengthened. Through text messaging teens are able to tell their parents they love them, the extended family is kept abreast with one other, and migrant women can mother their children from afar. Although Filipinos are culturally chatty, emotions and feelings are especially difficult to convey verbally. Confronting a person of the opposite sex, or person presumed to be superior (a parent, a boss, the wealthy, the religious clergy) is difficult and embarrassing. Text messaging has given Filipinos a venue to overcome *hiya* and text what is difficult or awkward to say. As Ellwood-Clayton points out, 'texting is uprooting traditional courtship, and providing young Filipinas with a new site in which they can experiment with romantic agency' (2005, p. 214). Filipinos' desire always to be in the know stems from curiosity about the world around them and the day-to-day lives of family and friends, coupled with distrust of government and media resulting from years of colonization and oppressive governments. Through text messaging, users are able to keep abreast with the goings on in both the private and public spheres. News and information are gathered and verified through a network of acquaintances, friends, and family. The cell phone thus operates like a modern-day town crier or evening paper.

Cell Phone Technology and Mothering

New media technologies 'are embedded in everyday life and its domestic and urban environments ... permeating all the mundane activities' (Lister *et al.*, 2003, pp. 219–220). Everyday life—'the family relationships, routines, cultural practices and spaces through which people make sense of the world' (Lister *et al.*, 2003, p. 220)—is central to the study of media and culture, but is often neglected in the study of new media technologies. It is acknowledged that new media have transformed or will transform the way we live and conduct our daily lives, but what exactly 'the nature of this transformation [means] is contentious: for some observers new media offer new creativities and possibilities, for others they reinforce and extend existing social constraints and power relationships' (Lister *et al.*, 2003, p. 220). The status of mobile communication technologies has shifted from being an exclusive communication device to a status symbol to something commonplace.

Although many of the Filipina domestic workers who seek employment abroad have some form of college education, their salaries are among the lowest among OFWs. They are paid by the hour and do not receive health or social security benefits from their employers. Computers and Internet connections are too expensive, so they rely on the cell phone to communicate with their families back home, sending text messages many times a day. According to David Morley, 'traditional ideas of home, homeland, and nation have been destabilized, both by new patterns of physical mobility and migration and by new communication technologies' (2000, p. 3). For the migrant mothers interviewed for this project, home is two places—the Philippines, where their children and spouses are, and their host country, where daily patterns and new lives have been created.

According to Morley, technology has 'also worked to break up the traditional family structures around which the idea of home has been based' (2000, p. 26). As Trisha, who has been living apart from her children for six years, says:

> It wasn't even a big deal when my green card arrived. As long as you are earning income and you are at peace with the lifestyle with your husband, then you don't worry. You get used to it. Besides there are so many ways to keep in touch.

Thus the transnational household becomes normal and naturalized into daily life. When he needs money, 21-year-old Victor Morillo texts his mother, a fabric cutter in Bahrain, to wire some over (Diamond, 2002). A few days later, funds would be wired to a local bank. His mother is able to provide for him whenever he needs it despite the fact she is in another country. This system of dependence that is expected within a family is expanded to exist in the global world through communication technologies.

Children of migrant parents need 'concrete reinforcement of their parents' love' and that can be achieved by 'creating "bridges" of communication' (Parreñas, 2003, pp. 141–142). Through text messaging, mothers in far-away countries are still able to 'mother' their kids by asking them how their day was and whether school went okay, reminding them to eat dinner and do their homework, and bidding them goodnight almost as if they were there. Rosario Reyes, the Filipina domestic worker killed by a suicide bomb in Israel, transmitted a message to her son the night before her death: 'Go to sleep' (Diamond, 2002). Donna, a nanny in New York, has three sons in the Philippines. Although she has not seen them for almost two years, she keeps in constant contact—sending over 20 text messages daily. She says, 'I'm able to find out what they are doing, how they are. It's like I am there.'

Annabel, also a nanny in New York, has a two-year-old son in the Philippines and has been away from him for a year. While he is too young to use a cell phone, she is constantly sending text messages to her husband. Like Donna, she sends text messages many times a day. 'I probably text an average of thirty texts a day', Annabel says, laughing, 'and my bill is about $60 on text messaging alone, plus $80 for calls each month. Sometimes it's more than that.' Although she spends a large portion of her hard-earned dollars on keeping in touch with her husband, she says she has no choice because it is what keeps her sane. She says: 'I ask them what they are doing, what did the baby do today, what did they eat. Those are our conversations. And then we text sweet (romantic) things to each other. It's better than nothing.' This is what Pertierra calls an 'absent presence', where cell phones are used not only to connect but also 'mediate and shape relationships' (2005, p. 26). Those apart are able to maintain presence and build associations despite physical distance.

For Juliet who has two children, a 17-year-old daughter and a 20-year-old son, text messaging allows for more productive weekly phone calls:

> I text them messages and instructions as they come to me so that when I talk to them on the weekend, they already have the answers I need. You know how it is when you're on the phone, the conversation goes all over the place. But because I text them ahead, the things we have to talk about is not forgotten.

Both Juliet and Donna maintain a Philippines cell phone number where their families send them text messages that only cost US2¢ versus the US24¢ to send a text message to an international cell phone number. They then send their replies through their US cell phone, which costs about 15¢ per text. There is no charge for incoming text messages or calls on Philippine cell phones so arrangements like this allow migrant mothers to curb costs.

During an afternoon I spent with Donna, I noticed she was punching away on her cell phone and making verbal comments about the messages she was receiving. When asked what the matter was, she replied:

> This crazy son of mine is asking permission to borrow the car. He sends me a text saying that his dad says its okay and to just let me know. But my husband texted me to tell me that he told our son that we would still talk about it.

What struck me about this exchange was that Donna was halfway across the world, had not seen her teenage son for almost two years, and yet wielded the power to give or withhold permission for the use of the family car. Mundane, daily family interactions such as this were being conducted via text message, allowing her to maintain her position as mother in the family.

According to Lister *et al.* 'the consumption of media technologies is both shaped by and shapes existing family dynamics' (2003, p. 236). As mother, Donna is perceived to have the 'natural' right to monitor and control her children's behaviour and activities; something that most migrant parents are unable to maintain. Through text messaging, her position in the family as the parent who 'gives permission' is upheld and sustained, despite physical absence. In addition, her dual role as nurturer and provider is strengthened, thus making her the more dominant parental figure despite the distance. Donna herself says that her husband defers to her in making major family decisions, even if he is the one living with and physically caring for the children because she is the one who earns the money. This is in contrast to traditional roles where although 'it is often the case that men hand their wages over to their wives, who in turn manage and distribute household income, this by no means implies that women are able to take major decisions' (Chant & McIlwaine, 1995, p. 9). Her role as primary breadwinner and decision maker for the family can be understood as the masculinization of motherhood, especially in the context of a traditional Filipino family where women are often relegated to domestic concerns.

According to Morley, 'communication technologies can function as disembedding mechanisms, powerfully enabling individuals (and sometimes whole families or communities) to escape, at least imaginatively from their geographical locations' (2000, pp. 149–150). The instantaneity and mobility of cell phones and text messaging allow users to 'forget' or momentarily suspend their physical separation. This is clearly the case for Donna:

> It's like I am at home. I worry about the same things that I would worry about if I were with them—what they had for dinner, whether they did their homework, did they go grocery shopping—those are the questions I text, they are the same questions I would ask when we were all together.

Thus, text messaging allows her to straddle two worlds at the same time 'forgetting' that distance and time zones separate her from her loved ones. Because the content of her text messages is mundane and about everyday concerns, she is able to 'escape' from the reality of the separation, allowing her to 'enjoy an absent presence' (Pertierra, 2005, p. 42).

Technology is often seen as a negative outcome of a capitalistic society that alienates people from one another. As we have seen, for migrant Filipino mothers, technology can be empowering and humanizing for those far away from loved ones. Participation, albeit virtual and mediated, can exist despite the separation of space and time. For the migrant mother, technology helps with the dislocation brought about by moving to a foreign land. Being in the same physical space with loved ones is, of course, irreplaceable. Yet for those in situations where it is not possible, the virtual space enabled by communication technologies bridges the physical gap. According to Meyrowitz: 'when we communicate through telephone, radio, television, or computer, where we are physically no longer determines where and who we are socially' (1985, p. 115). Overseas Filipino workers no longer have to be in the same location to participate within assigned social roles: technology makes physical presence less of a necessity.

For Hondagneu-Sotelo and Avila, transnational mothering involves 'forsaking deeply felt beliefs that biological mothers should raise their own children', leaving them instead to be cared for by the extended family (1997, p. 557). When Annabel left for the United States, care for her then one-year-old son was entrusted to her mother, even if her husband lives in the same house. Yet in examining the text messages that Filipina women send to their children, we see traditional notions of motherhood being carried out. Although Donna's husband does not work, he spends his day shuttling his three sons to and from school, taking over some of the roles Donna used to occupy. Yet while Donna now has the authority to give permission and has the final say in family decisions, she is still concerned with what the kids had for breakfast, if they did their homework, and so on, thus still operating within the framework of a patriarchal society. This concern for her children's physical and emotional welfare is often rooted in guilt for being away and the belief that these are what 'good mothers' do. Although they are far away, migrant mothers are still responsible for nurturing and caring for the children. Parreñas points out that while migrant women 'experience a certain degree of gender liberation because of their greater contribution to household income ... even with these advances, patriarchy is not eliminated but is somehow retained in migration' (2003, p. 30). Established traditional gender roles are not reconstituted when mothers migrate abroad for work. While reinventing the concept of transnational motherhood through real-time and constant communication, cell phones have also paved the way for traditional ways of mothering to flourish.

Is it Really Empowering?

In listening to these mothers' stories, and in looking at the literature on new media technologies, we see that cell phones have given users a sense of empowerment. Indeed, cell phones have given these Filipino migrant mothers the ability to keep in touch and maintain or rebuild familial relationships and reassert their roles as mothers. Yet this same technology that empowers them to maintain their roles in the family also emphasizes the distance and reminds them that they are away and in many ways powerless. As the migrant mother leaves work for home at the end of the day, it is painfully real to her that it is time for her to wake her children up for school halfway across the globe. Although the migrant mother may feel powerless, communicating by cell phone technology can be read as a form of resistance for these migrant mothers against the dislocation they experience and the structures that have placed them in this position of powerlessness.

For Foucault, resistance 'can only exist in the field of power relations. But this does not mean they are only a reaction or rebound, forming with respect to the basic domination as underside that is always passive, doomed to perpetual defeat' (1978, p. 96). For Foucault, power is relational: that is, resistance and repression coexist with each other. Points of resistance are thus always present, something with which Parreñas concurs:

> Even though the forms of resistance that migrant Filipina domestic workers deploy against dislocations do not involve the direct diminishment of structural or institutional power, their acts of resistance must be credited with possible interventions against the way that structural inequalities operate in shaping their everyday lives. (2003, p. 34)

The cell phone and text messaging gives migrant mothers a way of negotiating the dislocation and alienation that working abroad as a domestic worker brings. It also allows her to bridge the distance between the family in the Philippines and life in the United States. For the OFW, text messaging remains not merely a form of repression, enforcing traditional familial roles, but acts as a form of resistance, opening the possibility to provide economic support and emotional care for the family. In identifying this as an act of resistance, we must also remember what Foucault notes: 'discourses are tactical elements or blocks operating in the field of force relations' (1978, pp. 101–102). As discourse comes from the same source, the same field of force, it can shift sides—the discourse of resistance can become the discourse of repression. For Filipino migrant mothers, text messaging presents a counter-discourse to the dominant one of separation and alienation from the family: 'there can exist different and even contradictory discourses within the same strategy; they can, on the contrary, circulate without changing their form from one strategy to another, opposing strategy' (Foucault, 1978, p. 102). This contradictory discourse of empowerment and masculinization is still part of the structure of oppression limiting Filipino migrant mothers to the traditional role of motherhood. In addition, the cell phone technology that empowers migrant women further entraps them in the web of transnational migrant labour, perpetuating the divide between the global North and South.

Of course, individuals are not simply pawns of social structures but neither is power simply about the will of individuals, a point made by Giddens, who shares with Foucault the view that power is relational. According to Giddens' theory of structuration, 'the structural properties of social systems are both the medium and the outcome of the practices that constitute those systems' (Giddens, 1979, p. 69). For Giddens, acts of agency and social structures are linked together, and it is through the repetition of individual acts of agency that structures are reproduced or altered (Gauntlett, 2002, p. 93). Accordingly, we can see that as Filipino women leave employment abroad and maintain transnational families via cell phone, the global labour system is reproduced and reinforced. Thus, the individual actions of each migrant woman to empower herself through text messaging perpetuates the system that marginalizes and oppresses her.

The Philippine economy is very dependent on the remittances sent by overseas workers, and thus agencies such as the Philippine Overseas Employment Agency, which monitors departing workers, and the Overseas Workers Welfare Administration (OWWA), which supports workers' rights in the receiving nations, have been created. Both these offices are under the jurisdiction of the Philippine Department of Labor and Employment. When migrant workers return home, especially at Christmas time, they are greeted with parties and festivities in local shopping malls thrown by the government. A special line at immigration at the airport is reserved for 'balikbayan' (literally meaning 'return to country') workers to ease their re-entry. When departing the country, overseas workers are exempted from travel tax and dependants are entitled to reduced rates. The Philippine government's honouring of these workers as 'new heroes', coupled with the proliferation of cell phone advertisements showing happy families text messaging each other from all parts of the globe, promotes the transnational family with the mother working abroad as normal and acceptable, presenting and advocating a counter-discourse to the dominant discourse of separation and abandonment.

Innovations in cell phone services have also made the transfer of money across the globe easier. Globe Telecom, one of the largest cell phone networks in the Philippines, has a service called 'G-Cash' that allows OFWs to send small amounts of cash home, bypassing disproportionate bank transfer fees. Globe publicized, for example, how Filipino workers abroad can send home up to $200 through licensed distributors in their host countries for a small processing fee, usually a fraction of bank fees (http://myglobe.com.ph). The recipient immediately receives a text message that money has been sent. The cash can then be picked up at any Globe Telecoms office or licensed partners at local Philippine shopping malls. This has been beneficial for both parties. For the sender, transfer fees are less, small amounts can be sent, and the cash is received quicker. For the recipient, processing fees are minimal, funds can be obtained in the evening and on weekends when banks are closed, and cash is instantaneous. For the migrant mother who is paid an hourly wage, this is a cost-efficient way to send money home, especially in times of need.

Globe Telecom's chief rival Smart Communications has launched a promotion called 'International Budget Text' where for a small processing fee of less than US50¢, subscribers in the Philippines can buy promotional packages of international text messages for a fixed period of time at a much lower cost (http://smart.com.ph). Instead of getting charged US30¢ per international text message sent, for a limited period of time subscribers to this promotion pay only about 1¢ for each message sent internationally to the United States or Hong Kong. For example, US40¢ buys a package of 20 US-bound international text messages that need to be consumed within

24 hours. While 20 text messages seems like a lot, for the average Filipino cell phone user this is easily consumed in a day. Thus for children of migrant workers, promotional packages like these are ideal, especially for special occasions such as birthdays or Christmas when they miss their mothers the most.

New media technologies such as the cell phone have thus made leaving for another country for employment more palatable and endurable. It is undeniable that each woman's individual experiences of text messaging her children multiple times a day gives her a sense of empowerment, but it also simultaneously disempowers her by ensuring that she is a productive link in the chain of migrant labour. As Parreñas says, 'agency is conditioned and therefore limited by the social processes from which it emerges and takes place. This means that resistance, as it recuperates power, does not necessarily bring positive change' (2003, p. 34). Thus, there is no real empowerment. Although migrant mothers experience a sense of empowerment in maintaining their familial roles through cell phones and text messaging, they are also reinforcing existing social systems that limit their choices. This same technology that empowers also ties women back both to the domestic sphere and to the political economic structures they cannot escape.

Conclusion

Undoubtedly, new media technologies such as the cell phone and text messaging have given migrant Filipino mothers a way of maintaining relationships with their children and asserting their roles in the family. The capacity of text messaging for real-time communication has allowed both profound and mundane 'conversations' to take place, despite the distance of time and space. This has empowered Filipina migrant workers to reassert their roles as mothers despite being physically absent in the day-to-day lives of their children. Cell phone technology has made physical presence less necessary in maintaining a family structure.

However, cell phones and text messaging have also thrust migrant mothers back to the traditional roles of mothering. While being the breadwinner and primary economic provider for the family has masculinized the mothering role of these migrant women, text messaging has also reinstated and reinvigorated traditional mothering roles of caring and nurturing. Rather than rearranging gender roles in Filipino households, cell phone technology reinforces traditional roles in a patriarchal system.

Constant communication, while easing the pain of separation, does not change the political-economic structures that have pushed these women to leave home. While their use of cell phones and text messaging to maintain the family can be read as acts of resistance and as experiences of empowerment, migrant mothers are further caught in a system of power in which they are helpless. The institutional support and structures provided by the Philippine government promote migration as an act of heroism. Philippine cell phone companies provide products, services, and advertisements to alleviate the hardships that these acts of heroism entail. Migration and separation from loved ones are thus naturalized and accepted because new media technology allows the transnational family to exist and transnational mothering to thrive.

Acknowledgements

The author would like to thank Sarah and Derek Sweetman, Rachel Martin, and Mark Escaler for invaluable feedback, comments, and support.

References

Arnold, W. (2000) 'Manila's talk of the town isn't talk at all', *New York Times*, 5 Jul., p. C1.

Banko Sentral ng Pilipinas (n.d.) Available at: http://www.bsp.gov.ph (accessed 11 Dec. 2005).

Central Intelligence Agency (CIA) (2006) *The World Factbook*, Central Intelligence Agency, Washington, DC, Available at: https://www.cia.gov/cia/publications/factbook/index.html (accessed 20 Oct. 2006).

Chant, S. & McIlwaine, C. (1995) *Women of a Lesser Cost: Female Labour, Foreign Exchange & Philippine Development*, Pluto Press, London and East Haven, CT.

Diamond, D. (2002) 'One nation, overseas', *Wired*, vol. 10, no. 2, Available at: http://www.wired.com/wired/archive/10.06/philippines.html (accessed 12 Jun. 2002).

Ellwood-Clayton, B. (2005) 'Desire and loathing in the cyber Philippines', in *The Inside Text: Social, Cultural and Design Perspectives on SMS*, eds R. Harper, L. Palen & A. Taylor, Springer, Dordrecht, pp. 195–219.

Foucault, M. (1978) *History of Sexuality: an Introduction*, reprinted 1990, Vintage Books, New York.

Gauntlett, D. (2002) *Media, Gender and Identity: an Introduction*, Routledge, London and New York.

Giddens, A. (1979) *Central Problems in Social Theory*, University of California Press, Berkeley.

Goggin, G. (2006) *Cell Phone Culture: Mobile Technology in Everyday Life*, Routledge, London and New York.

Hondagneu-Sotelo, P. & Avila, E. (1997) '"I'm here, but I'm there": the meanings of Latina transnational motherhood', *Gender & Society*, vol. 11, no. 5, pp. 548–571.

Levinson, P. (2004) *Cell Phone: the Story of the World's Most Mobile Medium and How it has Transformed Everything*, Palgrave Macmillan, New York.

Lister, M., Dovey, J., Giddings, S., Grant, I. & Kelly, K. (2003) *New Media: a Critical Introduction*, Routledge, London and New York.

Meyrowitz, J. (1985) *No Sense of Place: the Impact of Electronic Media on Social Behavior*, Oxford University Press, New York.

Morley, D. (2000) *Home Territories: Media, Mobility, and Identity*, Routledge, London and New York.

Parreñas, R. S. (2003) *Servants of Globalization: Women, Migration, and Domestic Work*, Ateneo de Manila University Press, Quezon City, Philippines.

Parreñas, R. S. (2005) *Children of Global Migration: Transnational Families and Gendered Woes*, Stanford University Press, Stanford, CA.

Pertierra, R. (2005) 'Mobile phones, identity and discursive intimacy', *Human Technology*, vol. 1, no. 1, pp. 23–44.

Pertierra, R., Ugarte, E. F., Pinggol, A., Hernandez, J. & Dacanay, N. L. (2002) *Txt-ing Selves: Cellphones and Philippine Modernity*, De La Salle University Press, Manila, Available at: http://www.finlandembassy.ph/texting1.htm

Philippine Overseas Employment Agency (POEA) (2004) *Annual Report*, Available at: http://www.poea.gov.ph (accessed 11 Dec. 2005).

Uy-Tioco, C. (2004) *Cell Phones as a Cultural Technology: New Ways of Communicating in the Philippines* MA Thesis, New School University, New York.

Williams, R. (1974) *Television: Technology and Cultural Form*, reprinted 1992, Wesleyan University Press, Hanover, NH.

Socio-cultural Aspects of Mobile Communication Technologies in Asia and the Pacific: a Discussion of the Recent Literature

Mark McLelland

David Gauntlett, in his foreword to *Japanese Cybercultures*, notes the tendency in the anglophone West to 'assume that people in other countries, using other languages, are probably doing things with Internet technology that are *pretty similar* to those applications we are familiar with' (Gauntlett, 2003, p. xii; emphasis in the original). However, as that collection goes on to make clear, particularly regarding mobile Internet applications, this is not always the case.

As a researcher with a background in Japanese studies, in an attempt to help undergraduate students think critically about the interrelationship between society, culture and (new) technologies, I try to encourage them to consider that technologies have a history and that the meanings underlying their deployment are highly culture specific. Until recently the (now ubiquitous and quotidian) mobile phone was a helpful example to use. However, this particular technological innovation is becoming increasingly difficult to debate in class, since few students can remember a time before mobile phones and not one grew up in a household (as I did) where the landline telephone was a seldom-used late arrival. (My mother first rented one solely for 'emergencies', of which we thankfully had precious few. Accordingly, that original phone set lasted my family for 20 years.)

My students look at me with blank expressions when I point out that at the end of the nineteenth century, for instance, the telephone was regarded by people in 'society' as a proletarian device, to be used by servants to telephone orders to florists and department stores; or that Proust seldom used it (preferring along with other educated people to communicate via letter). Indeed, in *In Search of Lost Time*, although the narrator occasionally uses the telephone to call up the theatre and listen to a live performance of an opera or a play (a now forgotten service offered to early subscribers), it would never have occurred to him to have a personal conversation via such an uncouth medium. (Just imagine Marcel thinking 'I'll give Albertine a ring and see what's she's up to'.) Flash forward to today's permanently wired 'socialite' to understand how our definitions of 'society' and the modes of communication considered appropriate between its members have changed. This example, as well as others from Proust, including his astonishment at the time/space compression brought about by new transport mechanisms such as the railroad and the automobile, is useful in underlining the mutually constitutive relationships that exist between 'the everyday consciousness of ordinary people' and the technologies that surround them (see Berger *et al.*, 1973).

I also try to get my students to think about the particularities of different technologies, how they signify in different geographical locations and among different social groups, through introducing case studies from other societies in our region. The mobile phone is one gadget that always catches their attention. However, despite some new and important work that has recently been published about Japanese mobile communications, we are still awaiting a good general overview of mobile communications (and other related 'new media') in the Asia-Pacific region that does not constantly refer the reader back to the primacy of European or North American models. Likewise there is no overview that attempts to theorize regional influences and flows such as the immense impact that Japanese and Korean technologies have had on the Chinese market as well as the developing markets of Southeast Asia.

Why is this so? Castells *et al.* (2004), in *The Mobile Communication Society: a Cross-cultural Analysis of Available Evidence on the Social Uses of Wireless Communication Technology*, the broadest comparative survey of mobile communications media so far available, notes the many difficulties involved in producing a useful overview of new mobile technologies on a global scale. The variety of languages in which mobile communication takes place and in which research is undertaken is obviously formidable, but even more so is the pace at which the development and spread of these devices is taking place; it is almost impossible to keep up. Indeed, as Castells *et al.* point out, although it is clear to researchers that the uptake of these technologies will have profound and diverse effects in different regions, 'which kind of effects, under which conditions, for whom and for what is an open question' (Castells *et al.*, 2004, p. 2). These are obviously complex and nuanced issues that require detailed field work and case studies—still sadly lacking for the majority of Asian societies (with the notable exception of Ito *et al.*'s excellent collection on Japan, to be discussed in more detail below).

Given the expensive and time-consuming nature of such research, there is a tendency for researchers to offer data about the penetration rate of new technologies and the demographics of their use as well as descriptions of the technologies themselves and relevant business and government policies. However, as Castells *et al.*

point out, 'the diffusion of wireless communication proceeds so fast that purely descriptive data rapidly becomes obsolete' (Castells *et al.*, 2004, p. 2) and it is for this reason that Rao and Mendoza's *Asia Unplugged: the Wireless and Mobile Media Boom in the Asia-Pacific*, despite having the most seemingly useful title, is the least helpful of recent publications in terms of teaching about mobile communications in the region. The editors of the collection have brought together over 21 authors who are largely industry specialists and journalists rather than academics and the discussion seems to be targeted mainly at a business and technology audience. Although the chapters offer details about what technology is available where, analyses of evolving regional and country-specific markets and services, and assessments of the impact that government policies have had on the unequal distribution of mobile media throughout the region, there is too little analysis of the very specific socio-economic contexts that have led to divergent and contrasting uptake of mobile communications in different parts of the region.

The volume is subtitled a *handbook* and this is the best way to use it—as a resource to dip into for quick, data-rich surveys of particular countries since the density of data and number of charts and tables makes it a difficult read for those looking for a more discursive introduction to the issues. A cultural and media studies audience would find the discussion to be lacking in the amount of attention paid to social and cultural factors. Much of the language is hyperbolic; we read of the 'broadband tidal wave phenomenon' that is sweeping South Korea, a country which has pioneered mobile handsets of 'astounding' sophistication—but no sense is given of why (other than proactive government policies) the Korean population should have been so receptive to this technology as consumers. The treatment of the widespread popularity of mobile phone use among young Japanese consumers is similarly lacking in insight. The book's chapters are full of relevant and useful statistics but the cultural background that has led to certain patterns of usage in one society and a different pattern of usage in others is not investigated.

We need to look elsewhere for answers to these questions and in this brief overview I propose to introduce some recent sources that offer a range of essays, papers and reports that might help us gauge some of the regional similarities as well as specific differences in the uptake of mobile communications in societies throughout the Asia-Pacific region over the last decade. The main drawback of this selection is that the 'Asia' that is represented (as in much of the other literature that attempts to look beyond European or North American paradigms) tends to refer primarily to Japan, Korea, China, and, to a lesser extent, the Philippines. Australia, too, falls off the map in these collections, despite being one of the most wired societies in the wider region.

Of all these texts, Castells *et al.*'s *The Mobile Communication Society* is the most useful in setting up the basic research approach, an approach which is sensitive to local contexts and developments and that tries not to prioritize 'Western' models. Indeed, as the authors of the report point out: 'Our emphasis on a cross-cultural approach comes essentially from our intention to avoid cultural ethnocentrism in building the argument' (Castells *et al.*, 2004, p. 2). Castells *et al.* survey mobile uptake in Europe,

North America and Pacific Asia, 'the three areas of the world where wireless communication is most developed (save Australia)' (Castells *et al.*, 2004, p. 2) and are aware that their report overlooks Africa, South America and large parts of Asia. However, based on the regions that are considered, the authors pinpoint four very general trends that characterize the uptake of mobile communications. These are, firstly, the 'deep connection' between wireless communication and the development of a 'mobile youth culture'; secondly, the transformation of language via texting and multimodality; thirdly, the impact of wireless communication on socio-political activism; and, finally, changes in the experience of time and space. These trends are indeed apparent in many of the studies of Asia-Pacific societies offered in the texts under discussion but, as might be expected, there are no identifiable multi-regional 'Asian' characteristics, each trend being inflected by very specific local factors.

Regarding the uptake of mobile communications, Castells *et al.* note that 'The United States lags behind Europe and Japan and Japan is ahead of Europe in the uses of the wireless Internet' (Castells *et al.*, 2004, p. 2). Indeed, 'A broad consensus has been formed that the usage of *keitai* [mobile phones], including especially the myriad wireless Internet applications, is central to the transformation of the Japanese information society', and that this usage represents 'a process significantly distinct from the development of the computer-based Internet in other countries' (Castells *et al.*, 2004, p. 103). It is no surprise, then, for researchers looking for paradigms of both Internet and mobile communications development that disrupt normative US and European patterns, that it is usually Japanese examples that are invoked. Although researchers working on Japanese society have long been wary of claims of Japanese exceptionalism, as far as mobile phones are concerned, there is a clear body of evidence that shows how this particular technology has developed in Japan in a manner quite divergent and distinct from mobile telephony in other regions.

While not 'unique' to Japan (since Japanese patterns of usage are similar in some respects to South Korea—although far less attention has been paid to this important regional centre), it is clear that pre-existing Japanese cultural norms and practices have exerted a strong influence on the development of *keitai* technology and its deployments. In Ito *et al.*'s *Personal, Portable, Pedestrian: Mobile Phones in Japanese Life*, so far the only comprehensive analysis of mobile telecommunications in Japan, the editors are right to insist on referring to Japanese cell/mobile phones by their Japanese moniker, *keitai*, in order to underline the cultural specificity of this technology, its design, functions, uses and meanings—emphasizing its role as 'an artifact located in a specific national context' (Matsuda, 2005, p. 37).

This important collection takes us beyond earlier hagiographic accounts of the development of mobile communications in Japan such as Matsunaga's (2001) *Birth of i-Mode* and Beck and Wade's (2002) *DoCoMo: Japan's Wireless Tsunami* which focus more on business models, policy developments, and technical and management issues. Ito *et al.*'s collection is significant for scholarship into mobile media in the Asia-Pacific in two ways: the contributors challenge the model that sees the United States as 'the supposed vanguard of the information society' (Ito, 2005, p. 3) while simultaneously dislocating the hegemony of anglophone media theory by drawing upon extensive Japanese-language scholarship. The editors set up the intellectual project of the volume by stressing the need to understand how cultural and social factors impact not only upon the take-up of a technology but also on its functions and design.

Ito points out that technologies are not universal; rather, it is necessary to attend to 'the heterogeneous co-constitution of technology across a transnational stage' (Ito, 2005, p. 7), and the contributors to the volume do a good job in outlining how Japanese society comprises very different 'technoscapes' that differ markedly from US and European locations. In Japan (and to an extent in other Asian societies), unlike the Internet, mobile telephony was not 'conceived by an elite and noncommercial technological priesthood and disseminated to the masses' (Ito, 2005, p. 9) but emerged out of Japanese consumers' love of 'gadget fetishism and technofashion', and the market was driven not by a business elite but grew out of the existing pager culture of Japanese teenage girls. This collection emphasizes the fact that although the Japanese *keitai* supports functions associated with landline telephones, it cannot be understood solely through models that assume the primacy of voice contact. Matsuda notes how the *keitai* is a 'technofashion gadget' (Matsuda, 2005, p. 20) driven by young consumers, especially young women.

In this respect, the Japanese *keitai* has much in common with mobile phones in Korea and China where the design of the phones themselves, as well as the manner in which they are accessorized (via downloads, ringtones, wallpaper, as well as cases, handles, straps and other add-ons) has become an important signifier of fashionable status and identity (Castells *et al.*, 2005, p. 50). There are differences, however: in Japan and Korea, the 'customization' of the phone itself is more likely to take the form (among women in particular) of 'cute' downloads and add-ons such as animated and plastic cartoon characters (Hjorth, 2005). In China, where the technology is less ubiquitous and therefore more a signifier of status, ornaments are likely to feature real precious metal and gem stones. However, as Katz & Sugiyama (2005) point out, the use of gem stones, in the Chinese context, is not reducible simply to ostentation but needs to be considered in a cultural context where certain stones symbolize 'esteem, good fortune, peace and love'—as indeed do the phone numbers themselves, which are chosen on account of their auspicious sound or combination.

Other kinds of customization are popular, too, elsewhere in Asia. Bell, for instance, notes the penchant for painted handsets popular among Malaysian youth. She mentions her bewilderment at the choices on offer in a local mall: designs featuring the face of Osama bin Laden and a plane crashing into the World Trade Center, hanging next to Hello Kitty and characters from Disney. As she notes, cell phones and their decorations clearly 'function as a kind of extended body politic' (Bell, 2005, p. 81).

The need to pay close attention to the local contexts of use is further emphasized by Bell in her chapter on 'mobile technologies from Asia' in Glotz *et al.*'s (2005) *Thumb Culture: the Meaning of Mobile Phones for Society* (a title that would make more sense if 'culture', 'meaning' and 'society' were in the plural). Bell quite rightly draws attention to the 'cultures of mobility in Asia' (at last the plural) when she suggests that 'what it means to be "mobile" . . . has distinct cultural meanings' (Bell, 2005, p. 70). Bell is also correct in stressing that the real impetus behind mobile communications development and uptake is Asia, and not the United States, which 'in many ways . . . is more the anomaly than the rule' (Bell, 2005, p. 70). Bell's chapter results from a larger pan-Asian research project and it is to be hoped that the results can be brought together in a monograph.

Although most of the research into mobile communications in Asia focuses upon Japan, Korea and China, there is some excellent work being done on the Philippines, and the chapters on that nation are among the best in Ling and Pedersen's (2005) *Mobile Communications*, Kim's (2005) *When Mobile Came* and the 2005 *Proceedings of the International Conference on Mobile Communication and Asian Modernities.* Goggin (2006), too, in *Cell-phone Culture*, is clearly intrigued by the dynamic uptake of mobile communications (particularly messaging) in the Philippines, and his detailed analysis of Philippine usage (as well as his critique of the sometimes utopian tone of existing research) is a welcome addition to a discussion which otherwise tends to draw its examples of non-Western trajectories from the more 'advanced' region of Northeast Asia.

Despite the economic challenges faced by most of the population, the mobile phone has become, according to Pertierra (2005, p. 157), 'the most successful technology to be introduced to the Philippines'. In less than a decade cell phones have massively outstripped the number of landlines (30 million cell phones to 7 million landlines), reaching a penetration rate of nearly 35 per cent. While the mobile uptake was originally driven by the middle-class business market, providers have targeted the lower end of the economic spectrum by offering free or inexpensive texting and prepay phone cards, allowing those with no credit history or fixed abode to access the technology (Castells *et al.*, 2004, p. 124).

In the Philippine context, the implications of this sudden access to a technology previously reserved for elites can be described as revolutionary. Goggin (2006, pp. 77–80), for instance, emphasizes the degree of agency afforded by mobile telephony to otherwise disenfranchised individuals who used instant messaging to disseminate anti-government and anti-authoritarian information, parodies and jokes. The cumulative effect of this e-protest was the 'coup de text' that overthrew the corrupt President Estrada in 2001. While instant messaging was, of course, crucial in organizing the massive street demonstrations that eventually led to the president's resignation, these demonstrations could not have come about except through the establishment of a culture of parody and dissent that had already been established via the mass circulation of anti-Estrada text messages.

Importantly, research being developed by Yu (2006) also emphasizes the significance of instant messaging in a society like China, whose media are under strict state control. Information about the SARS epidemic, for instance, which the government tried to suppress, was widely distributed via text messages. Furthermore, the government's increasingly absurd denials of the seriousness of the health scare led to the circulation of messages of a parodic and darkly humorous nature, thus further undermining any respect that the populace had for its leaders. The impact of mobile communications on generating and supporting a culture of resistance to corrupt and authoritarian states in the region is clearly a topic needing further investigation.

Another key research topic should be the way in which mobile communications enhance the ability of diasporic populations to keep in contact with those left behind in the homeland. As Bell argues, mobile communications in the Asia-Pacific region (and beyond) are crucial in 're-charting social relationships fractured by colonial and post-colonial geographic separations' (Bell, 2005, p. 69). The importance of mobile communications to Filipinos/as, for instance, is not limited to the borders of that nation but has implications for the wider region, given the Philippine's large population of migrant workers. Pertierra observes how migrant workers had previously experienced difficulty making telephone contact with family and friends remaining at home (Pertierra, 2005, p. 160; see also Paragas, 2005a, p. 189), a point also emphasized by Thomson (2005) who offers a detailed case study of the mobile options available to migrant workers in Singapore. Thus, while in richer countries of Asia, mobile telephony is often a complement to other kinds of fixed-line and broadband communication, for Filipinos/as at home and abroad it is an important substitute (Paragas, 2005b, p. 118).

However, although Pertierra notes that communication via mobile phones (especially texting) is convenient because of its easy access and relative cheapness, he suggests that the true appeal of the technology in the Philippines (and in Asia more widely) lies in the mobile phone's 'capacity to provide private and unsurveilled communication' (Pertierra, 2005, p. 157) away from the prying eyes of kinship networks, many members of which share the same limited living space. In this constraining environment, texting clearly offers a level of privacy impossible for vocal communication. Bearing this in mind, I sometimes give my students a quick quiz on mobile communications throughout the world and one question concerns which nationality sends the most SMS messages. No one ever gets it right: it is in fact the Philippines, with an average of eight messages per user per day. (This exceeds even Japan, another nation where text messages exceed voice calls, especially among young people (Okada, 2005, p. 49).)

Pertierra argues that young people (particularly women) are often particularly constrained in their behaviour—both while in the home and under parental surveillance, and in public where they are subject to stricter rules of decorum than those faced by men. This includes behaviour whilst on the telephone (especially in public spaces) which Paragas notes 'is a riddle in decorum that seems to be more confounding than in the west' (Paragas, 2005b, p. 119). As a consequence of this lack of private space, it is difficult for young people to negotiate erotic relationships and so texting has been taken up for this purpose, enabling 'the exploration and development of sexual agency' (Pertierra, 2005, p. 169).

Intriguingly, Angel Lin (2005) makes a similar observation regarding the deployment of SMS messages by migrant workers in southern China who use their phones' messaging service 'for purposes of dating and courtship'. However, in the Chinese case, the use of SMS for this purpose is not connected to lack of personal space so much as it is to Chinese conventions of courtship which involve the exchange of romantic epithets and proverbs which can be selected and downloaded from a fixed repertoire (particularly attractive to migrant workers with limited education and cultural capital). As Bell notes, the condensed, character-based nature of the Chinese script, which is more space efficient than the alphabet, offers 'a far richer messaging experience' (Bell, 2005, p. 68) since it enables more complex content to be displayed on a small screen.

The particularities of space in the Philippine context are also taken up by Paragas, who notes how mobile phone use on public transport is constrained by numerous factors, not least issues to do with personal safety. In the metro Philippines where a range of public transport is available across a broad price range (which determines the type of clientele that uses it), Paragas argues that 'modes of transport bear strongly upon mobile telephony' (Paragas, 2005b, p. 123). In the more expensive, air-conditioned transport options where commuters are insulated from the noise and bustle of the street, etiquette demands that spoken conversations via the mobile phone are kept short and quiet and that discussions of a personal nature be avoided. In cheaper, more crowded (and less uptight) travel environments, however, the audacious display of a new model of phone can draw attention and mark the owner out as a target for petty theft, leading to 'the fear the gadget may invite harm' (Paragas, 2005b, p. 122). Thus, in certain environments in the Philippines, Paragas notes how the commonsense notion that mobile phone ownership offers the owner a 'security blanket' is turned on its head as phone owners are afraid to take out and use the devices lest the phone becomes 'a threat to one's security in public transport' (Paragas, 2005b, p. 127). Here is a clear instance of how mobile phone use is variously negotiated in a variety of spaces on a daily basis. This discussion of space and the cultural expectations conditioning behaviour (according to age, gender, social class, etc.) is a useful heuristic to get students thinking about their own deployment of mobile communications technologies as they navigate their way through public and private spaces in the (sub)urban landscape.

An interesting point of contrast regarding public transport and the mobile phone is offered by Japan, where spoken communication via these devices on buses and trains is frowned upon and needs to be handled with particular decorum. As Matsuda (2005, p. 24) points out 'The physical noise is not the problem. Rather, *keitai* conversations disrupt the order of urban space' through confusing the boundaries of private and public. Okabe & Ito (2005) point out how from 1996, when young people became the main demographic to take up the mobile phone, a voluble media discourse developed about their use in public settings. (Indeed, in Japan, switching one's phone to vibrate/silent is referred to as 'manner mode'.) Unlike the more chaotic nature of the cheaper forms of public transport in the Philippines, public transport in Japan is 'characterized by precise technical and social regulation and very low rates of disorder' (Okabe & Ito, 2005, p. 208) supported by commuters' willing acquiescence to a system of 'mutual surveillance, regulation and sanctioning' (Okabe & Ito, 2005, p. 208). (It must, however, be noted that 'Japan' is not uniform in this respect and that commuters in the Osaka area are more forgiving of *keitai* use by commuters than those in Tokyo.)

As Castells *et al.* (2004) point out, mobile communications clearly impact upon and themselves help to redefine notions of time and place. That communication via mobile devices itself creates a new kind of virtual environment is clear from many case studies and, as Castells *et al.* also point out, it has often been young people who have pioneered and shaped these environments. The importance of understanding the particular developmental trajectories of new media is highlighted by another Japanese case—the popularity in the mid- to late 1990s of paging devices among high-school girls (a mobile communications phase absent in the youth cultures of Australia, Europe and the United States). Okada (2005) notes how in the mid-1990s schoolgirls in Japan appropriated paging devices originally developed for the (male) business market, using the keypad on telephones to send simple text messages to the LCD displays of friends' pagers. This led to a distinct *poke-kotoba* (pager-lingo) among the subculture and assisted young people in organizing their private lives outside of parental supervision and control. The prior popularity of pagers among youth groups impacted upon the functions of early models of mobile phones which included SMS services and, from 1999, via a service known as *i-mode*, Internet enabled e-mail (Kohiyama, 2005). In Japan, these were innovations driven by and for the youth market.

The above examples bear witness to the four general trends apparent in mobile communications uptake outlined by Castells *et al.* That many markets in the region are youth-driven is clearly apparent (and of particular interest is the extent to which young women have been attracted to mobile technologies). Language change, too, which has always been influenced by youth subcultures, is also clearly being driven by this youth uptake. The political ramifications of the kind of mass networking enabled by instant messaging is also clearly demonstrated in the case of the popular revolt against corruption in the Philippines, but there are other less apparent (for the moment) changes taking place among the new Chinese middle class whose irreverent attitude towards their ineffectual local and national leaders is increasingly expressed via instant messaging. That people's consciousness of time and space is also being affected by new developments in mobile communications is also clear from small-scale local differences in the use of mobile devices in public spaces (especially on public transport) to wider patterns of adoption among mobile, diasporic populations.

Although, as the texts discussed above illustrate, we now have many interesting and useful individual articles investigating these different dimensions of mobile communications and their impacts, the field still lacks a general overview that attempts to understand these patterns and trends in a regional framework, especially one that does not fall back upon default references to a standardized or normalized pattern of usage in 'the West' (almost always understood to be the United States or Europe, never Australia, the particularity of whose mobile uptake remains largely unexamined). Such a review is necessary to help encourage students (and some professional researchers) to think of mobile communications (or indeed any technology) as part of complex socio-cultural systems that are tied into local, regional as well as global flows of people, technology and desire.

References

Beck, J. & Wade, M. (2002) *DoCoMo: Japan's Wireless Tsunami*, American Management Association, New York.

Bell, G. (2005) 'The age of the thumb: a cultural reading of mobile technologies from Asia', in *Thumb Culture: the Meaning of Mobile Phones for Society*, eds P. Glotz, S. Bertschi & C. Locke, Transcript Press, Piscataway, NJ, pp. 67–88.

Berger, P., Berger, B. & Kellner, H. (1973) *The Homeless Mind: Modernization and Consciousness*, Penguin, Harmondsworth.

Castells, M., Fernandez-Ardevol, M., Qiu, J. & Sey, A. (2004) *The Mobile Communication Society: a Cross-cultural Analysis of Available Evidence on the Social Uses of Wireless Communication Technology*, Annenberg Research Network on International Communication, Annenberg.

Gauntlett, D. (2003) 'Foreword', in *Japanese Cybercultures*, eds N. Gottlieb & M. McLelland, Routledge, London, pp. xii–xv.

Glotz, P., Bertschi, S. & Locke, C. (eds) (2005) *Thumb Culture: the Meaning of Mobile Phones for Society*, Transcript Press, Piscataway, NJ.

Goggin, G. (2006) *Cell-phone Culture: Mobile Technology in Everyday Life*, Routledge, London.

Hjorth, L. (2005) 'Locating the mobile: customisation and gender in the Asia-Pacific region', in *Proceedings of the International Conference on Mobile Communication and Asian Modernities*, City University of Hong Kong, Hong Kong, 7–8 Jun., pp. 119–140.

Ito, M., Okabe, D. & Matsuda, M. (eds) (2005) in *Personal, Portable, Pedestrian: Mobile Phones in Japanese Life*, eds M. Ito, D. Okabe & M. Matsuda, MIT Press, Cambridge, MA.

Ito, M. (2005) 'Introduction: personal, portable, pedestrian', in *Personal, Portable, Pedestrian: Mobile Phones in Japanese Life*, eds M. Ito, D. Okabe & M. Matsuda, MIT Press, Cambridge, MA, pp. 2–16.

Katz, J. & Sugiyama, S. (2005) 'Mobile phones as fashion statements: the co-creation of mobile communication's public meaning', in *Mobile Communications: Re-negotiation of the Social Sphere*, eds R. Ling & P. Pedersen, Springer, London, pp. 63–82.

Kim, S. (ed.) (2005) *When Mobile Came: the Cultural and Social Impact of Mobile Communication*, Communication Books, Seoul.

Kohiyama, K. (2005) 'A decade in the development of mobile communications in Japan (1993–2002)', in *Personal, Portable, Pedestrian: Mobile Phones in Japanese Life*, eds M. Ito, D. Okabe & M. Matsuda, MIT Press, Cambridge, MA, pp. 61–76.

Lin, A. (2005) 'Romance and sexual ideologies in SMS manuals circulating among migrant workers in southern China', in *Proceedings of the International Conference on Mobile Communication and Asian Modernities*, City University of Hong Kong, Hong Kong, 7–8 Jun., pp. 141–156.

Ling, R. & Pedersen, P. (eds) (2005) *Mobile Communications: Re-negotiation of the Social Sphere*, Springer, London.

Matsuda, M. (2005) 'Discourses of keitai in Japan', in *Personal, Portable, Pedestrian: Mobile Phones in Japanese Life*, eds M. Ito, D. Okabe & M. Matsuda, MIT Press, Cambridge, MA, pp. 19–39.

Matsunaga, M. (2001) *The Birth of i-Mode*, Chuang-Yi Publishing, Singapore.

Okabe, D. & Ito, M. (2005) 'Keitai in public transportation', in *Personal, Portable, Pedestrian: Mobile Phones in Japanese Life*, eds M. Ito, D. Okabe & M. Matsuda, MIT Press, Cambridge, MA, pp. 205–218.

Okada, T. (2005) 'Youth culture and the shaping of Japanese mobile media: personalization and the keitai Internet as multimedia', in *Personal, Portable, Pedestrian: Mobile Phones in Japanese Life*, eds M. Ito, D. Okabe & M. Matsuda, MIT Press, Cambridge, MA, pp. 41–60.

Paragas, F. (2005a) 'A case study on the continuum of landline and mobile phone services in the Philippines', in *When Mobile Came: the Cultural and Social Impact of Mobile Communication*, ed. S. Kim, Communication Books, Seoul, pp. 178–197.

Paragas, F. (2005b) 'Being mobile with the mobile: cellular telephony and renegotiations of public transport as public sphere', in *Mobile Communications: Re-negotiation of the Social Sphere*, eds R. Ling & P. Pedersen, Springer, London, pp. 113–130.

Pertierra, R. (2005) 'Without a room of your own? Get a cell phone', in *Proceedings of the International Conference on Mobile Communication and Asian Modernities*, City University of Hong Kong, Hong Kong, 7–8 Jun., pp. 157–170.

Proceedings of the International Conference on Mobile Communication and Asian Modernities, City University of Hong Kong, Hong Kong, 7–8 Jun. 2005.

Rao, M. & Mendoza, L. (eds) (2005) *Asia Unplugged: the Wireless and Mobile Media Boom in the Asia-Pacific: the Asia-Pacific Internet Handbook, Episode VI*, Response Books, New Delhi.

Thomson, E. (2005) 'Reaching home by hand phone: foreign worker communities and mobile communication in Singapore', in *Proceedings of the International Conference on Mobile Communication and Asian Modernities*, City University of Hong Kong, Hong Kong, 7–8 Jun., pp. 199–213.

Yu, H. (2006) 'Talking, linking, clicking: the politics of AIDS and SARS in urban China', Paper presented at the seminar 'Understanding the Internet in the Asia-Pacific', University of Wollongong, 25–26 Sept., Abstract available at: http://www.capstrans.edu.au/resources/conferences/2006/25sep-wkshp-prog-bios.pdf (accessed 29 December 2006).

Picture This: the Impact of Mobile Camera Phones on Personal Photographic Practices

Lisa Gye

The convergence of the camera and mobile phone has proved to be highly popular. This should come as no surprise to anyone interested in both the history of mobility and the history of photography.[1] As media archaeologist Errki Huhtamo has asserted, the first mobile medium proper was arguably amateur or personal photography (Huhtamo, 2004). Camera phones are not, however, just another kind of camera. Located as they are in a device that is not only connected to the telecommunications grid but that is usually carried with us wherever we go, camera phones are both extending existing personal imaging practices and allowing for the evolution of new kinds of imaging practices. Given the centrality of personal photography to processes of identity formation and memorialization, changes to the ways in which we capture, store, and disseminate personal photographs through the use of devices like camera phones will have important repercussions for how we understand who we are and how we remember the past. The aim of this paper is to look at the current social uses of personal photography and to consider the impact that camera phones will have on these uses. I will examine the ways in which camera phones are enabling new modes of personal photography, which will extend the role that photographs play in our lives.

The Social Uses of Personal Photography

In order to understand the relationship between existing personal photographic practices and new practices emerging from camera phone usage, we need to understand why people take photographs in the first place. Surprisingly, despite its tremendous popularity as a practice, personal photography has received relatively scant attention from scholars. (There are a few important studies in this field. See, for example, Sontag, 1978; Bourdieu, 1990; Barthes, 1980; Chalfen, 1987; Hirsch, 1997). One study (Van House *et al.*, 2004) does provide us with a useful set of categories that can help us to understand the reasons why people take personal photographs; in order to construct personal and group memory; in order to create and maintain social relationships; and for the purposes of self-expression and self-presentation. I will first examine each of these categories as they relate to older analogue photographic practices before looking at the ways in which these categories apply to new practices emerging out of camera phone culture.

Constructing Personal and Group Memory

The introduction of celluloid film and easy-to-use box cameras in the late 1880s sparked what Huhtamo describes as a 'camera epidemic' in Victorian society (Huhtamo, 2004, website). But this epidemic was far from spontaneous. As Pierre Bourdieu observed in his now famous sociological study of 1965, *Photography: a Middle-brow Art*, the *desire* to photograph is not a given—it is socially constructed and culturally specific. Writers who have tracked the evolution of personal photography seem to concur that the rise in its popularity can be directly attributable to the emergence of a correlation in the public imagination between photographic practice and private memorialization. Leading this change in public perception was Kodak Eastman—the premier manufacturer of portable photographic devices in the late nineteenth and early twentieth century. As Liberty Walton notes:

> By the turn of the century, Kodak no longer promoted the camera's instantaneous capabilities that were a novelty in the 1888 promotions. Instead, the idea of the snapshot's value as an aid to memory was promoted. The idea that photography could be used to capture and save moments is evident in Eastman's advertising campaign, containing such slogans as '... a means of keeping green the Christmas memories.' 1903: 'A vacation without a Kodak is a vacation wasted.' 1904: 'Where there's a child, there should the Kodak be. As a means of keeping green the Christmas memories, or as a gift, it's a holiday delight.' 1905: 'Bring your Vacation Home in a Kodak.' 1907: 'In every home there's a story for the Kodak to record—not merely a travel story and the story of summer holidays, but the story of Christmas, of the winter evening gathering and of the house party.' 1909: 'There are Kodak stories everywhere'. (Walton, 2002, pp. 26–38)

Personal snapshot photography took off on the back of such rhetoric. As Bourdieu notes, as 'a private technique, photography manufactures private images of private life ... Apart from a tiny minority of aesthetes, photographers see the recording of family life as the primary function of photography' (1990, p. 30). Evidence of the importance of photography to private memorialization is all around us—in our own collections and albums. It can also be heard in the often-told story of survivors of natural (or unnatural) disasters—after securing the children and the pets, one must also ensure that the family photographs are rescued as well.

Personal photographs not only bind us to our own pasts—they bind us to the pasts of those social groups to which we belong. The documentation of social groups through photography reinforces our connections to others. Photographs are often the sutures that bind the narratives of group memory. As Giuliana Bruno argues:

> In a post-modern age, memories are no longer Proustian madeleines, but photographs. The past has become a collection of photographic, filmic or televisual images. We, like the replicants [of *Blade Runner*], are put in the position of reclaiming a history by means of its reproduction. (Bruno, 1990, p. 183)

Creating and Maintaining Social Relationships

'Photos reflect social relationships but they also help to construct and maintain them' (Van House *et al.*, 2004, p. 7). Exchanging and sharing personal photographs is integral for the maintenance of relationships. One important function of personal photography, one that extends its existence as a material prosthesis for personal memory, is the role it plays as an aid to storytelling. Sharing memories through the creation of narratives around them plays an integral role in the construction and maintenance of personal relationships. As Richard Chalfen argues, personal photography is 'primarily a medium of communication' (Chalfen, 1991, p. 5).

People share personal photographs in many ways. They send copies of photographs through the mail (and now e-mail) to distant loved ones. They frame the photographs for display in their homes and workplaces. They attach them to pinboards and fridges. They post them to photosharing sites such as Flickr (http://www.flickr.com), Photobucket (http://photobucket.com) or Kodakgallery (http://www.kodakgallery.com). However, while in the strictest sense of the word these activities involve sharing, in reality sharing requires some dialogue and annotation. These activities emphasize display and can operate without any dialogue or annotation at all. But when they do function in silence, their ability to build and maintain social relationships is somewhat diminished.

Photographic albums, on the other hand, function by virtue of presence. Not only do they make the subjects of the photographs present to us in the form of their image, as Barthes famously noted, but they require, if they are to enact their primary function, the *presence* of a narrator and a listener (see Barthes, 1980). The performative act of showing and telling is integral to the ritualized use of personal photography. Not that this, as we all know, is always a pleasant experience. As Martha Langford points out:

> Looking at another person's snapshots, slides, home movies or tapes can indeed be killing: presentations are rarely of short duration, and repetition seems endemic to the genre. The real-life domestic experience is loaded with compensatory pleasures—intimacy, conviviality, emotional investment and perhaps a slice of cake. Inside stories frame the pictures, animating even the most stilted of studio portraits with family secrets and subversive tales. (Langford, 2001, p. 5)

Our preference for face-to-face sharing, coupled with our resistance to annotation, as demonstrated in research undertaken by Nancy Van House and others, has implications for digital photography generally and more specifically for mobile photography (see, in particular, Van House *et al.*, 2004). This will be discussed in more detail below. For now it is enough to note that photography is integral to the creation and maintenance of social relationships and that these relationships are reinforced by activities such as the narration of stories that accompany face-to-face photosharing.

Self-expression and Self-presentation

As an expressive medium, photography has been both celebrated and condemned. While artistic photography has had to fight against the prejudices of the art establishment to become accepted as a legitimate artistic practice, personal photography has been more heartily embraced as an important mode of self-expression. As Susan Sontag notes: '[n]obody takes the same picture of the same thing', so 'photographs are evidence not only of what's there but of what an individual sees, not just a record but an evaluation of the world' (Sontag, 1978, p. 88).

Individuals are engaged at all levels of personal photographic production—from the taking of the snapshots and now, with digital photography, to their processing and dissemination. The emphasis on user control is a long-standing feature of personal photography of all kinds. Ease of use was a key selling point for Kodak Eastman in the early twentieth century and remains so for all kinds of cameras today. Kodak Eastman advertising assured its customers that anyone could be a photographer—even women and children! (for examples of the 'Eastman version', see Collins, 1990).

What do the photographs we take tell the world about who we are? Presumably, that our view of the world is unique and interesting and that, by virtue of this, so are we. At least that is what we hope. This is related but different from the use of personal photography for self-presentation. Self-presentation relates more to those photographs we take or display of ourselves, our family, our friends, our possessions, our pets, and so on. Photographs which are taken or used for self-presentation reflect the view of our selves that we want to project out into the world.

The Impact of Camera Phones on the Social Uses of Personal Photography

Camera phones participate in the same kind of economy of photography as that just discussed. They can be used to take photographs that help to construct personal and group memories, that maintain and develop social relationships and that allow their users to express and present themselves to the world. However, there are certain affordances which are specific to the camera phone that mean that these functions are somewhat altered and that allow for new forms of photographic practice. By examining each of these categories separately we may be able to discern the ways in which camera phones reinforce and extend existing photographic practices and the ways in which they create new practices.

Constructing Personal and Group Memory

The fact that we tend to carry our mobile camera phones with us wherever we go means that we now have increased opportunities for taking photographs. As Dong-Hoo Lee notes in her study of the use of camera phones by women in Korea, 'with camera phones, taking pictures is experienced as an everyday activity' (Lee, 2005, website). However, there are certain features of mobile imaging that have prevented the camera phone thus far from becoming the imaging device of choice at important moments in our lives that we may wish to memorialize. The poor resolution of camera phone images has, until recently, meant that significant life events such as weddings, births and so on are still photographed using cameras rather than camera phones. As many images from events such as these end up as prints that can be archived or shared, image resolution is important.

Mobile camera phones suffer from similar problems to digital cameras—the very immaterial nature of the technology works against our usual ways of working with personal photographs. If personal photographs operate as a medium of communication, enabling shared conversations and storytelling, then digital camera technologies work against some of the enabling techniques of this practice. For instance, sharing photographs, even when they are in albums, is often a tactile and sensual experience. Photographs are passed between participants who are invited to inspect them more closely. As Sit *et al.* note:

> These forms of interaction are well supported with printed photographs in communal spaces, such as gatherings around the kitchen table or the living room sofa. In contrast, this naturalness of interaction has not been duplicated with digitally formatted photos published online. (Sit *et al.*, 2005, p. 1)

Creating the same kinds of interaction with digital images on mobile devices is made even more difficult by their location in a highly personalized device. While we may show photos to people on our mobile devices, we are usually reluctant to hand the device over to someone who may not understand the interface and so we keep them at a distance from the images under examination.

The construction of personal and group memories through photography is inscribed within an oral tradition which, ironically given its primary function, is not facilitated as yet by the mobile camera phone. Efforts are currently being made to allow users to annotate their camera photos and send them to online repositories or directly to family and friends. Nokia's Lifeblog offers users a commercial and domesticated version of the moblog—as Gerard Goggin points out, Nokia itself carefully avoids the term 'moblog' because of its association with technical digital culture, preferring instead to use phrases such as 'mobile sharing' and 'life sharing' (Goggin, 2005). Nokia is keen to promote Lifeblog as a family-friendly platform and its advertising emphasizes the family connection. Expounding the benefits to consumers, Nokia's publicists write that:

> Nokia Lifeblog provides a simple method of capturing your daily experiences and unforgettable moments, like a child's birth or a friend's wedding and storing them all in one place. The memories you want to share can be easily posted to the web. The blogs can be accessed by the family, friends and colleagues via a password-protected area, or can be available for general access. (Nokia, 2005)

However, even annotated Lifeblogs are unable to promote the kinds of two-way conversations that are integral to the popularity and endurance of personal photography as a social practice for the construction of personal and group memory.

Creating and Maintaining Social Relationships

While any form of photosharing is bound to assist in the creation and maintenance of social relationships, the kinds of photos that are most often taken with mobile camera phones are those that reinforce the user's individuality rather than their ties to other groups. Research conducted by Daisuke Okabe and Mizuko Ito clearly supports this idea. They argue that:

> the social function of the camera phone differs from the social function of the camera in some important ways. In comparison to the traditional camera, most of the images taken by camera phone are short-lived and ephemeral. The camera phone is a more ubiquitous and lightweight presence, and is used for more personal, less objectified viewpoint and sharing among intimates. Traditionally, the camera would get trotted out for special excursions and events—noteworthy moments bracketed off from the mundane ... The camera phone tends to be used more frequently as a kind of archive of a personal trajectory or viewpoint on the world, a collection of fragments of everyday life. (Okabe & Ito, 2006)

This supports Daniel Palmer's argument that the Nokia moment is far more intimate than the Kodak moment (Palmer, 2005). It is also reflected in the kinds of images that are captured using mobile camera phones. Okabe and Ito have observed that '[w]ithin the broader ecology of personal record-keeping and archiving technologies, camera phone images occupy a niche that is more personal, fleeting, and commonplace' (Okabe & Ito, 2006). Camera phones and other mobile multimedia, according to Ilpo Koskinen, tend to participate in aesthetics of banality. That is, the images captured with these devices tend to focus on the mundane, trivial aspects of everyday life. Koskinen argues that '[p]eople capture ordinary things in immediate life and share them with friends and acquaintances in monadic clusters that become even more emotionally and relationally more self-reliant than before' (Koskinen, 2005, p. 15).

Other studies have pointed to a distinction between captured camera phone images that are affective and captured camera phone images that are functional (Kindberg *et al.*, 2005). Under the taxonomy that they developed as a result of their research, Kindberg *et al.* argue that affective images can be taken to enhance a mutual experience or to share an experience with someone who is absent, either in the moment or later. Affective images can also be taken to be used for personal reflection, rather than to share with others. Interestingly, almost half of the photos taken by the participants in their study fell into this category, something which supports Okabe and Ito's assertion that the camera phone is primarily a personal imaging device. Similarly, functional images could be divided into three categories: images that support a mutual task, images that serve as a reminder either to the self or others to perform a remote task, or images that serve as a personal reminder to perform a practical, individual task.

Regardless of whether images being captured on camera phones are affective or functional, most researchers studying camera phone use agree that the images themselves are far more individualized, mundane and everyday than much of the personal photography that preceded it.

While the images may be more individualized, the connectivity of camera phones presents us with new ways to maintain our social relationships through photosharing. Instantaneous photosharing can lead to a practice that is peculiar to mobile camera phones and which Ito describes as 'intimate, visual co-presence' (Ito, 2005). Just as text messaging has allowed people to remain in perpetual contact, Ito argues that by keeping in touch through picture messaging camera phone users are able to create a shared visual space—a sense of presence created through visual intimacy. Mark Federman takes this argument one step further by claiming that:

> One of the most important effects of massively multi-way, instantaneous and ubiquitous communications is pervasive proximity. We experience everyone to whom we are connected—and conceivably everyone to whom we are potentially connected—as if they are exactly next to us. The effect is that of hundreds, or thousands, or millions of people coming together in zero space, so that there is no perceptible distance between them. (Federman, 2006)

As appealing as this may seem, reality suggests that this kind of scenario is not widespread. In fact, sharing images through services such as MMS (Multimedia Messaging Services) has not taken off as yet in most parts of the world. In fact, much of the sharing that takes place with camera phones still takes place in face-to-face situations rather than through the sending of images. This is partly because of the cost of sending MMS. However it is also because users are unclear about such things as handset compatibility and how to send the images (see, for example, Kindberg *et al.*, 2005).

Self-expression and Self-presentation

While intimate, visual co-presence may yet be some way off, there is no doubt that the increasing pervasiveness of personal camera technologies, which are exemplified by the mobile camera phone, is leading to a change in the way in which we visualize the world. As Van House *et al.* argue: 'Ready access to imaging encourages people to see the world "photographically"—as images, and to see beauty and interest in the everyday. And easy internet-based sharing creates an audience' (Van House *et al.*, 2005).

Mobile camera phones certainly increase our ability to express ourselves and to present our unique view of the world to others. However, the transitory nature of camera phone images means that self expression is shifting away from 'this is what I saw then' to 'this is what I see now'. The photo message in particular participates in this economy of presence. As Mark Federman argues, '[p]hotographs, clippings, scrapbooks, and even works of art and sculpture as forms of cultural memory give way to ephemeral artefacts that exist for a brief instant in the span of time, as a sharing of experience itself' (Federman, 2006, p. 5).

Seeing the world photographically is also, as Susan Sontag points out, in itself somewhat problematic. As she reminds us, '[w]hile there appears to be nothing that photography can't devour, whatever can't be photographed becomes less important' (Movius, 1975). Translated to aspects of public life such as news and current affairs, the increasing use of mobile camera images by mainstream media organizations, which in itself is an interesting phenomenon, serves to remind us that the term 'newsworthy' has now become interchangeable with the term 'pictureworthy'. If an event is unable to be captured by an imaging device, its value is diminished. Hence the somewhat perverse desire that appears to have developed in Western cultures to document all aspects of our lives lest we are deemed not to exist.

This point leads us to ask what kinds of resistances have appeared to the use of such pervasive imaging devices. Despite its incredible popularity as a medium, misuses of photography and intrusive photographic practices have long been of concern to people with an interest in both personal privacy and the public sphere. Huhtamo reminds us that this goes back as far as the introduction of portable cameras in the late nineteenth century:

> Their activities developed into a kind of distributed panopticon—anybody anywhere could be the target of a snapshot. Caricaturists often interpreted such intrusions as sexually motivated (the pleasure beach being a favourite setting), but the camera also seemed to have a de-humanizing effect on the person carrying it. In a telling cartoon a group of ladies are seen pointing their cameras at a man hanging from a branch of a tree over water, struggling for his life. Instead of terror or empathy, the faces of the ladies show excitement about the photogenic event; as can be expected, none of them makes the slightest effort to run for rescue. (Huhtamo, 2004)

This distrust of the roaming photographer and his/her panoptic technology is resurfacing today in the current distrust of the mobile camera phone. While phone and camera manufacturers scramble to find ways to promote the use of mobile camera phones for the documentation of private life, public concern about the inappropriate use of these technologies is escalating. These two competing discourses are set to shape the evolution of the mobile camera phone and its attendant practices and software.

Finally, revisiting Koskinen's argument with regards to the banality of mobile imaging, one might also argue that the intensely intimate nature of camera phone imagery leads to a reinforcement of a movement towards what Richard Sennet has called the 'tyranny of intimacy' (Sennet, as quoted in Palmer, 2003). Drawing on Sennet, Palmer argues:

> We are permeated with narcissism, not as self-love but in terms of the exclusive reference to ourselves, which asks: What does this event mean to me? The narcissistic subject searches for and expects 'real' and intense experiences only within the framework of their own needs and expectations. (Palmer, 2005, p. 163)

By reinforcing the intensely personal, camera phones may also participate in this narrow economy of self, one that becomes almost asocial in its emphasis on the individual and its refusal of the more broadly social.

Conclusion

There is no doubt that mobile camera phones are having an impact on the established ways in which we record and archive our personal and group memories, create and maintain social relationships and express and present ourselves to our friends, family and the world. As camera resolutions improve and as users become more comfortable sharing their photographs through MMS, photosharing sites and camera technologies such as Bluetooth, we will undoubtedly see many more changes to the ways in which we undertake these activities. Research in this area is, at best, nascent. Solid empirical work needs to be undertaken if we are to really understand the current and potential impacts of camera phone technologies on our culture and our lives. For now we can say that camera phones are set to extend our way of looking at the world photographically and in doing so bring changes to how we understand ourselves and that world.

Note

[1] According to a published report by ABI Research, by 2008 more than one billion camera phones will be in service in markets throughout the world. While sales numbers in Australia are difficult to locate, analysts have noted that four years after Sharp and J-Phone introduced the first camera phone to the Japanese market, 85 per cent of handsets sold in Japan in 2005 have a built-in camera. In Korea, the number is closer to 99 per cent. See 'Mobile phone imaging: opportunities for driving usage of camera phones through click/send/print' (ABI Research, 2005, http://www.abiresearch.com/products/market_research/Mobile_Phone_Imaging), 'Camera phones: disruptive innovation for imaging' (Market and trend report, 5th Version of 11 Oct. 2004, http://www.eurotechnology.com/store/camera-phone/) and 'Camera phones to get 99% of local market in 2005—Korea' (*Camera Phone Reviews*, 22 Nov. 2004, http://www.livingroom.org.au/cameraphone/archives/camera_phones_to_get_99_of_local_market_in_2005_korea.php).

References

Barthes, R. (1980) *Camera Lucida: Reflections on Photography*, Random House, London.

Bourdieu, P. (1990) *Photography: a Middle-brow Art*, Polity Press, Cambridge.

Bruno, G. (1990) 'Ramble city: postmodernism and *Blade Runner*', in *Alien Zone*, ed. A. Kuhn, Verso Books, New York, pp. 183–195.

Chalfen, R. (1987) *Snapshot Versions of Life*, Bowling Green State University Popular Press, Bowling Green, OH.

Chalfen, R. (1991) *Turning Leaves: the Photographic Collections of Two Japanese American Families*, University of New Mexico Press, Albuquerque.

Collins, D. (1990) *The Story of Kodak*, Abrams, New York.

Federman, M. (2006) 'Memories of now', *Receiver*, 15, p. 4, Available at: http://www.receiver.vodafone.can/15/articles/pdf/15_08.pdf (accessed 23 Mar. 2007).

Goggin, G. (2005) '"Have fun and change the world": moblogging, mobile phone culture and the Internet', Paper delivered at *BlogTalk Downunder*, Sydney, 20 May, Available at: http://incsub.org/blogtalk/?page_id = 119 (accessed 23 Mar. 2007).

Hirsch, M. (1997) *Family Frames: Photography, Narrative, and Postmemory*, Harvard University Press, Cambridge, MA.

Huhtamo, E. (2004) 'An archaeology of mobile media', Keynote address, *ISEA*, Available at: http://www.isea2004.net/register/view_attachment.php?id = 6230.

Ito, M. (2005) 'Intimate visual co-presence', Position paper for the *Seventh International Conference on Ubiquitous Computing*, Tokyo, 11–14 September 2005, Available at: http://www.spasojevic.org/pics/PICS/ito.ubicompos.pdf (accessed 23 Mar. 2007).

Kindberg, T., Spasojevic, M., Fleck, R. & Sellen, A. (2005) 'The ubiquitous camera: an in-depth study of camera phone use', in *IEEE Pervasive Computing*, vol. 4, no. 2, Apr.–Jun.

Koskinen, I. (2005) 'Seeing with mobile images: towards perpetual visual contact', in *A Sense of Place: The Global and the Local in Mobile Communication*, ed. K. Nyiri, Passagen Verlag, Vienna.

Langford, M. (2001) *Suspended Conversations: the Afterlife of Memory in Photographic Albums*, McGill-Queen's University Press, Quebéc City.

Lee, D. (2005) 'Women's creation of camera phone culture', *Fibreculture Journal*, no. 6, Available at: http://journal.fibreculture.org/issue6/issue6_donghoo.html

Movius, G. (1975) 'An interview with Susan Sontag', *Boston Review*, Jun., Available at: http://bostonreview.net/BR01.1/sontag.html (accessed 23 Mar. 2007).

Nokia. (2005) 'Lifeblog backgrounder', Press document, Available at: http://www.nokia.com.

Okabe, D. & Ito, M. (2006) 'Everyday contexts of camera phone use: steps toward technosocial ethnographic frameworks', in *Mobile Communication in Everyday Life: an Ethnographic View*, eds J. Höflich & M. Hartmann, Frank & Timme, Berlin.

Palmer, D. (2003) 'The paradox of user control', in *Digital Art and Culture Melbourne 2003 Conference Proceedings*, RMIT, Melbourne, pp. 167–172.

Palmer, D. (2005) 'Mobile exchanges', Paper presented at the *Vital Signs* conference, ACMI, Melbourne, 8 Sept.

Sit, R. Y., Hollan, J. D. & Griswold, W. D. (2005) 'Digital photos as conversational anchors', *Proceedings of the 38th Hawaii International Conference on System Sciences*, Hawaii, 3–6 January 2005, IEEE Hawaii, Available at: http://ieexplore.ieee.org/iel5/9518/30166/01385459.pdf?amumber=1385459 (accessed 23 Mar. 2007).

Sontag, S. (1978) *On Photography*, Farrar, Straus & Giroux, New York.

Van House, N., Davis, M., Ames, M., Finn, M. & Viswanathan, V. (2005) 'The uses of personal networked digital imaging: an empirical study of cameraphone photos and sharing', Paper presented at *CHI 2005*, Portland, Oregon, 2–7 Apr.

Van House, N. A., Davis, M., Takhteyev, Y., Ames, M. & Finn, M. (2004) 'The social uses of personal photography: methods for projecting future imaging applications', Available at: http://www.sims.berkeley.edu/- ~ vanhouse/vanhouseetal2004b.pdf.

Walton, L. (2002) 'How shall I frame myself?', *British Columbia Historical News*, vol. 35, no. 4, pp. 26–38.

The Cameraphone and Online Image Sharing

Virginia Nightingale

Image Sharing and the Cameraphone

The cameraphone and the cell phone together create a commercially exploitable mobile communications environment, yet there is evidence that the mobile communications industries find encouraging a culture of online cameraphone image sharing difficult. According to M:Metrics (2006a), 'cameraphone ownership and usage is climbing across the U.S. and Europe', but it is unclear that this is in response to user demand. Rather, it reflects the now almost ubiquitous bundling of camera and cell phone. A user survey by Lyra Research of 525 cameraphone users in the United States, conducted in April/May 2006 indicated that 72 per cent of users share images 'on phone', 28 per cent use MMS to send images, 31 per cent send images using e-mail, and 33 per cent save images to their PC, but only 12 per cent share their images online. This suggest that the cameraphone is most used for highly personal activities that involve the least additional cost to the user, and that online sharing of images does not yet register as a mainstream component of mobile culture (Lyra, 2006). Unfortunately for cameraphone users, the industry has been slow to address the reasons for this lack of interest, and so only a few sites, most notably *Flickr*, have risen to the 'social networking' challenge. InfoTrends/CAP Ventures (2006) claims that the divergent interests of industry groups actively retard development of services that could release the potential of the mobile imaging market:

> Improvements in the cameraphone experience are occurring, but they have been relatively slow due to differing priorities of handset manufacturers, wireless carriers, and retail photofinishers. Concerns regarding costs, usage models, pricing, and revenue opportunities are affecting product design and service offerings to the detriment of maximizing the mobile imaging potential. (InfoTrends/CAP Ventures, 2006)

Mobile image archiving in effect justifies the bundling of the camera with the mobile phone by encouraging the production of a resource that is increasingly valuable in cyber culture: user-generated content (M:Metrics, 2006b). User-generated content attracts the digital general public, delivers consumers to advertisers, and provides rich information about consumers to both the communications industry and to advertisers. Yet the success of the image collections in securing ongoing commitment from site members is jeopardized by the poor quality of the experiences delivered at the site. Site features available are common to all image archives—e.g. the 'editor's pick of the day'; display of 'most recent' submissions on the site's homepage; lists of the images most commented on or most viewed; and comment and tagging options. These options, all borrowed from other online forms (blogs and forums in particular), make image archiving and online sharing more accessible, but also show that the industry is not tackling how the service might be improved.

Cameraphone Image Archives

Broadly speaking three types of sites offer opportunities to exhibit cameraphone images online: photojournals, moblogs, and image archives (Kato *et al.*, 2005, p. 306). In 2005 I monitored three cameraphone sites: TextAmerica; Foneblog; and Mobloguk. While these encouraged the development of a sharing ethos, comment and sharing took second place to image archiving. Users at the sites we monitored seemed to use the sites mostly for image storage, and as a result the sites struggled to capitalize on their capacity for entertainment and networking. Comments on images more usually addressed the subject of the image than the photographic practice.

By mid-2005, the sites included options for submitting cameraphone video materials. At about this time the moral concern associated with cameraphones shifted from protection of privacy (for example, snapping unsuspecting topless bathing beauties and sharing the photos online) to the moral outrage occasioned by the British phenomenon of 'happy slapping' (see Zimonjic & Hastings, 2005; Akwagyiram, 2005). Encouraged by the example and popularity of reality TV, young people created their own 'reality scripts' for the cameraphone by 'slapping' unsuspecting victims—a commuter snoozing on the bus; a person walking down the street—then sharing the video with a circle of friends and/or uploading it to the Internet. Whether justified or not, this panic inadvertently confirmed the entertainment value that could be extracted from the online sharing of cameraphone videos and images.

By late 2005 innovative software developers had created sites specifically for online entertainment that featured 'social networking' based on image sharing. *Flickr* had emerged as the leading host for cameraphone images and was acquired by Yahoo!, and YouTube had debuted online. The giant corporations of cyberspace—News Limited, Yahoo!, Microsoft, and Google—all reacted to the perceived potential commercial value of social networking and its capacity to attract user-generated content as entertainment. They began investing heavily in its development. However, the scale of the online presence of the cyber giants puts pressure on the dedicated cameraphone hosts (like TextAmerica and Foneblog), which seem increasingly endangered.

The cameraphone hosts diversified their activities. In 2005 we witnessed a shift from archiving to more playful community activities, involving guessing and mocking games, and the introduction of tagging (using technorati and buzznet). By 2006 the sites expanded their activities again. The games continued, but the sites had begun to encourage users to sort images into categories (albums), and now offer chat and (text-based) moblogging options. In spite of these enhancements, the sites lag behind the industry leaders (YouTube and MySpace). The industry has resorted to less than subtle persuasion to 'encourage' mobile image sharing. In 2005 manufacturers began to replace removable photo cards with non-removable image storage, thus pushing the use of mobile connectivity for online access and complicating the image-sharing process. In late 2005 Foneblog entered into a deal with the Irish telecommunications company O_2, to channel new cameraphone owners to the site for a free trial period. O_2 bundled 'free' Foneblog access for three months with its new phone contracts, though, of course, call charges still applied.

The mobility of the cameraphone increases the likelihood that it will be used for controversial purposes, and its versatility is increasingly attracting social and political concern, resulting in more rigorous site surveillance. The online user community is routinely expected to share the burden of site surveillance (e.g. the ubiquitous, 'If this comment is offensive please forward this email notification to: . . .'). The growth of advertising online has further intensified the prevalence of surveillance and monitoring. In mid-2006 Reiter's Camera Phone Report noted advertiser pressure on MySpace to rid itself of objectionable material. MySpace prevailed on other sites that help users post online content to assist, in particular Photobucket:

> Photobucket has 30 people working full time on monitoring images and will spend $1 million during the next year on the effort. The company can monitor more than half the images and wants to be able to check all of them.

Clearly user-driven surveillance is not working to the satisfaction of the advertising industry, so surveillance is metamorphosing into a new online industry. The sites need advertising to make a profit, and advertisers are attracted to the entertainment and interest that the presence of controversial material generates, but cannot risk their products being linked to content that might damage brand images. The response has been to further classify content and restrict access, but these activities are contrary to the sharing ethos on which the long-term success of such sites depends.

The Cameraphone's Networks of Exchange

Much has been said about digital media and the expansion of user agency in the online environment. Instead of being passive receivers of prefabricated and mass distributed entertainments, users are considered to have been liberated by new media because they have control simultaneously of the means to produce, distribute and receive media products. John Hartley (2005) has applied the term 'redaction' to the increased control users may now exercise over publication of their creative work, though as Sinnreich has commented:

> Although this new paradigm confers a long-sought agency upon a previously powerless class of individuals, it also raises the bar for participation, potentially

> widening the gap between information haves and have-nots, the media literate and illiterate. (Sinnreich, 2007, p. 45)

The recent popularity of mass user sites like Myspace and YouTube, the heavy industry investment in them, and the routine deployment of surveillance and monitoring, together suggest that any benefits of redaction could be offset by the exclusion of significant user constituencies, especially if provision of user-generated content becomes a precondition for engagement. For example, one of the advantages of YouTube over the dedicated cameraphone sites has been its capacity for recycling materials created by others (and by media industry professionals in particular) as opposed to creating the shared content oneself.

With the massive investments in recent site takeovers, the industry has opened itself up to dependence on the ongoing acquisition of high-quality user-generated content to make the sites profitable. In this sense the industry's investment can be seen as the first offering in a gift exchange—a *prestation* or *offering* to Internet users, even though clearly motivated by self-interest. In turn, this raises the question of why it is that users should reciprocate by willingly creating and freely bestowing content 'gifts' on the site owners, when they already pay for access to the technologies and services that make the gift exchange possible. To answer this it is obvious that we need to untangle the network of exchange revealed by the offering of gifts at such sites.

In 1925 Marcel Mauss published his seminal analysis of gift exchange in archaic societies (Mauss, 1966/1925). Mauss was motivated by a problem homologous to the one addressed here. Like other early anthropologists, Mauss was fascinated by the practice of the potlatch in some North American Indian tribes. The practice simultaneously links status and sacrificial giving—sacrificing goods to the point of personal and communal annihilation—thus demonstrating a belief that the gods/spirits will repay the gift in kind, and with interest. Of course it is those who see themselves as bound by the requirement to repay the generosity of the gift who actually repay—not only in kind but by conferring status and loyalty. Mauss recognized that there are parallels between the systems of exchange in tribal cultures and capitalist investment in risky ventures where the return is sought, not from the gods but from constituencies (like mass audiences). Social (and economic) control is in effect maintained by ties of reciprocity, and Mauss sought to identify the principles that govern the expectation that an apparently unsolicited gift received will be repaid. He wanted to understand 'what force there is in the thing given which compels the recipient to make a return.' His stated intention was:

> to isolate one important set of phenomena: namely, prestations which are in theory voluntary, disinterested and spontaneous, but are in fact obligatory and interested. The form usually taken is that of the gift generously offered; but the accompanying behaviour is formal pretence and social deception, while the transaction itself is based on obligation and economic self-interest. We shall note the various principles behind this necessary form of exchange which is nothing less than the division of labour itself. (Mauss, 1966/1925, p. 1)

Here Mauss makes the important connection between gift exchanges, a division of labour and the maintenance of social order. In the network of exchange that surrounds the cameraphone and its image archives we see similar principles at work. Focusing on the network of exchange allows us to situate very different understandings of cameraphone culture within the one cultural system. It draws attention to the activities of manufacturers and telecommunications companies as much as it does to site users. It explains the exclusion of users from industry activities as a component of the division of labour brought about by the initial entrepreneurial investment in the production of cameraphone devices.

Georgina Born has recently noted that anthropologist Alfred Gell (1998) was strongly influenced by Mauss's analysis, which he applied to the analysis of creative process and agency (Born, 2006, p. 15). Rather than limiting the study of art objects (images and archives in this case), to the objects themselves, Gell suggested a taxonomic approach to the networks of exchange that surround the production and deployment of cultural works. He pointed out that agency is an unstable phenomenon in the context of art because where the image is situated, how it is being used, and when in its life-cycle such questions are asked, all influence who is exercising agency and what is meant by their exercising that agency. Gell begins with the proposition that the actual content or meaning of any particular image or image making is less important than the social processes and relations on which creative production depends and that it serves. This basic proposition is also shared by Actor Network Theory.

In the cameraphone context, industry players maintain the ongoing operational environment and offer 'patronage' to site users. For Gell, patronage occurs when the creative work of the artist is directly controlled by the image prototype (the subject of the creative work, or in semiotic terms, the referent) (Gell, 1998, pp. 36–37). As patron, the prototype controls both the conditions under which the creative work is produced and the environment of reception in which the image is displayed. By describing this type of agency as patronage I have attempted to evoke a sense of both its pervasiveness of the social relations brought into being by the cameraphone sites and also the limitations it places on the 'agency' of the cameraphone user. This does not meant that the industry's patronage is repressive in terms of the user's creative and aesthetic expression—quite the opposite. But it does suggest caution in blanket assumptions that users are able to engage completely freely, as they might like, in such environments. Image hosting sites (including 'shock sites' or sites that specialize in images likely to offend the general public) maintain systems of surveillance and exercise the power to remove unwanted materials as soon as they are reported. All require that users satisfy a range of commercial and legal preconditions (e.g. mobile phone contracts, site user agreements, software user agreements, sometimes even ongoing direct debit agreements).

The Division of Labour

Figure 1 charts the key (though obviously not all) actions and players involved in the production and persistence of the online cameraphone experience.[1] At each

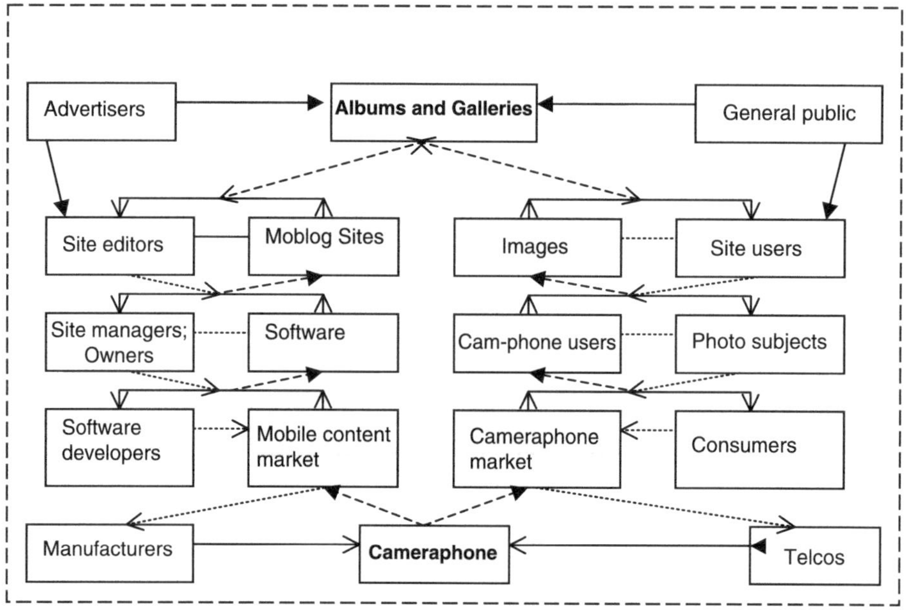

Figure 1 The cameraphone 'environment' and its division of labour.

(horizontal) level the outcome/product of the social relationship is identified, starting with the cameraphone and ending with the galleries and albums displayed on the websites. The shaded boxes denote the cultural 'objects' produced at each level. The cameraphone introduces a primary division of labour between those who maintain the system by paid labour and those who maintain it by levies (contractual arrangements) and by gifted labour (the offering of user-generated content). Each cultural 'object'/gift actively implicates new constituents, stakeholders and subjects in the system of exchange. So, for example, the camera phone market implicates consumers who are 'produced' as cameraphone users; cameraphone users implicate family members, friends, even total strangers in their cameraphone activities. The images produced as a result of this interaction work selectively on some of the cameraphone users who subsequently are motivated to share their images online.

Business-to-business transactions are separated off from the business-to-consumer transactions even though business-to-business transactions are, paradoxically, highly dependent on cameraphone usage and site visitation. Albums and galleries, like the cameraphone itself, occupy a primary site of intersection between the interests of industry and of users, and are therefore the spaces where industry and users interact. Initially, the cell phone manufacturer and the telco underwrite the cameraphone environment. Their entrepreneurial collaboration results in the creation of the *cameraphone* (a gift to the mobile communications environment), but it also leads to two other 'creations': a *market for cameraphones* and a *market for content*. The telcos and manufacturers are vulnerable to these markets and dependent on both the consuming public and skilled industry professionals to ensure a return on their investments.

At the moment of its birth as a commodity the cameraphone therefore activates a series of new and pre-existing social, economic and cultural relationships. The deployment of the technology in two different communities—the general public and the ICT industry—is visible as a cell-like division that defines the places where key actions and transformations take place. The market for cameraphones is managed primarily by the telcos, and it addresses the general public. The mobile content market, on the other hand, addresses software and site developers—people with specialized production skills and a vision, perhaps, for the future of mobile technologies. So from the moment of the cameraphone's invention, the general public is situated in a relation of reciprocation through consumption through phone usage and online image sharing. The cameraphone's attractiveness to consumers lies in its capacity to activate new and pre-existing social relations in interesting ways.

The *cameraphone market* generates industry revenue on the basis of the sale and use of cameraphones, while the *mobile content market* generates the conditions for a series of new entrepreneurial initiatives, and attracts additional social actors to the environment and to the development of user services. The new actors include potential site managers/editors or moderators, search service providers, advertisers, and companies involved in supplying an increasing array of imaging services. The new entrants effectively distribute the costs of the cameraphone project across more industry players. They service the needs of the sites for space, status and income, but they call into question the capacity of the sites to balance the needs of the new players alongside those of the site users and patrons. In other words, where the sites originally provided services to two constituencies (the site users and the manufacturers/telcos), they increasingly also provide services to additional constituencies whose needs may not be totally consistent with the needs of users, but who (like the digital general public), are attracted to the sites by the images and image archives. The advertising industry is one example.

The Cameraphone Experience

Industry control of the cameraphone environment means that user agency is best understood as *reciprocity* rather than as *response*. The 'recipients' of the industry 'gift' are not cameraphone users only but other industry actors as well. However, user agency may be expressed through image creation and/or site usage. It is constrained by the relations between image creator and image subject as much as by the technical specifications of the cameraphone or the user's ability. However, focusing only on images and image making to investigate user agency can be misleading if it focuses only on user-generated content or products and directs attention away from critical analysis of the industry-controlled 'environment'. On the other hand, emphasizing industry control of the cameraphone environment tends to underestimate the industry's dependence on users. In effect industry is less 'free' to walk away from the gift/investment than users. The division of labour ensures that the user has room to act with considerable autonomy within the sphere of action defined by the network of exchange. The network of exchange links the user and the industry most closely at the moment of consumption—when a cameraphone is purchased—and at the point where images are shared online.

For the user, however, the cameraphone experience is more about the images and what they can, or wish they could, do with them. From their perspective, the cameraphone experience cannot be separated from its role in life offline. As noted above, most cameraphone activity—'on phone' display; sending images by MMS or e-mail; saving messages to a PC—occurs outside the industry environment covered by Fig. 1. These cameraphone activities have become part of the pattern of exchange and reciprocity that make up the fabric of everyday life (Ito *et al.*, 2005; Ling, 2004). In this environment it has been shown that the cell phone is implicated in often quite complicated gift transactions. The user decides not only what photos to take but also whether and to whom a phone, phone camera or call credits may be loaned, and who may view the images taken. It is only in the activity of online sharing that the user interacts with the mobile communications industry and a global digital public where the expectation for the exchange of user-generated content for services becomes explicit. The cameraphone experience therefore encompasses the mobile but offline activities of mobile culture.

Mobile image sharing extends this everyday usage into the arena of online culture. It brings to the online sharing situation expectations based in the patterns of everyday usage. Fujimoto (2005, p. 80ff.) has argued that this culture is relatively independent of an online context since many mobile communications activities do not involve blogging. Instead, he proposes that the term *mobility* be expanded to distinguish between *mobility* and *mobilism*. Fujimoto limits *mobility* to the 'functional dimensions of portability and freedom from geographic and social constraint' (p. 80), and suggests the term, *mobilism*, for the 'broader cultural and social dimensions such as malleability, fluctuation and mobilization' (p. 80) that emerge as mobile communications are integrated into the flow of everyday life. *Mobilism* has a significance for mobile culture beyond the Japanese context which produced it. It offers clues for understanding the value that users attach to the specifically cameraphone experience, whatever the national contexts.

Mobilism implies independence—release from the need for a home base and even for a personal computer, and while the following examples reinstate the 'blog' for its importance as a way of organizing a personal online 'voice', they also draw attention to the different ways in which users enact *mobilism* as a form of online activism. The term 'activism' is used here to refer to the ways in which online image sharing needs a purpose, beyond archiving and online storage, that makes sense to the user. As we will see below, such purposes may be creative, personal or political, or combinations of all three.

Individual Cameraphone Websites

The introduction of the cameraphone initially provoked interest in 'authored' image archives—archives that are both 'art works' and 'authored' by solo enthusiasts. The Hand

Collector (see www.robottle.com/hands/), for example, explores a connection between the hand print and the elaborate digital displays of cyberspace. It is thematic—a collection of photos of painted, decorated or embellished hands. The Hand Collector plays a curatorial role from his (*sic*) base in Italy: collecting, selecting and displaying images submitted by a global constituency of contributors. Each image is authored—a name and country of origin provided. The images straddle the line between amateur and professional art/work, and double as online entertainment. The mobilism here is expressed as a global project, where the Hand Collector is a secondary patron (borrowing agency from the industry) to this living work.

By contrast, Mizuko Ito maintains an archive of cameraphone photos of *bento* (Japanese lunch boxes) on her website. Ito authors both the site and the images she creates and displays online. She has photographed hundreds of *bento* she has created. In addition, she provides links to other *bento* sites; to her personal website; to her academic writing (Ito *et al.*, 2005); and to online collections of *bento* recipes. Most interesting perhaps, Ito presents the *bento* archive as a reflection on its cultural politics and significance in Japanese everyday life (see also Fujimoto, 2005, p. 89). Ito's site ties cameraphone image making to the enactment of social relationships and ongoing ties of kinship and affection. She claims it as an intervention in gender politics. Her practice domesticates mobility—the cameraphone doesn't leave the kitchen—but the images do. So here we see *mobilism* without *mobility.* Online sharing allows the images and the lunchbox to escape into the realm of global communications. The activity is at once both domestic labour and engagement in national and global cultural expression. In addition, Ito has surrounded the image archive with additional textual material that outlines a range of social purposes served by the site: links to recipes, a source of inspiration, a meditation on Japanese life, an expression of motherly love, an intervention in gender politics. Her activism as site creator and artist is to re-present certain 'life genres' of Japanese life. She inscribes the *bento* into mobile culture, while simultaneously anchoring the Internet record of her daily creations. Ito argues that mobile communications knit online and offline experiences together and reduce online alienation (Ito, 2005, p. 8).

My third example is the well-documented case of Sydney's Man of Lettuce cablog (www.cablog.com.au; Goggin, 2005), a personal blog that combines *mobility* and *mobilism*. Neylan, the site's author, documents incidents and thoughts that arise in the flow of his work and life as a Sydney taxi driver. Unlike the previous examples that are static in place but mobile in time, this site is suffused with both. Agency is enacted on-the-move and in-the-world. An interesting aspect of Neylan's arrangement of his image archive is the way the photos evolve as a blog *about* cameraphone photography:

> Two shots of Port Macquarie at sundown. Couldn't decide between under exposure or normal exposure. Yours truly planting a commemorative boab tree last week for Steve Irwin. Raindrops on a deserted gravel road. Have a great long weekend. Drive carefully though, lots of traffic police early this morning enforcing the double-demerits points period which started at midnight. Go the Swanees ! *(forget the NRL)* (September 29, 2006)

Neylan repeats images to disrupt the reading process, so this entry was accompanied by four images—two of them the same except for a change in exposure. The second image of Port Macquarie, ostensibly taken to explore how best to use the camera, doubles as a commentary on the practice of cameraphone photography. It intrudes both textually and visually. Visual interruption in the flow of image reading is even more obvious in Neylan's image archive, where image repetition operates as a rhetorical form. Certain images in effect punctuate the visitor's reading of the album, and anchor the reader in places and spaces that document Neylan's moments of *im-mobility*: the moments when photos are taken. In this sense, Neylan's blog presents an innovative exploitation of the mobility inherent in his daily work as a taxi driver. Neylan's *mobilism* is expressed in text rather than image, and used to endorse national culture and local events. By referencing a series of events of current national (Australian) significance—the death of Steve Irwin; the dangers of driving on a holiday weekend; and an Australian Football League grand final—Neylan floats topics for possible blog discussion. These issues assume similar interests, activities and values among site visitors and are designed to mobilize a response from Australian cablog readers.

Ito and her associates argue that mobile phone activities ground the Internet in the practices of everyday life, and assist users to develop a sense of continuity between life online and off; that they are more significant for the personalized *mobilism* they support than for their *mobility*:

> The extroverted, out-of-doors nature of mobile communication, as well as its low-profile origins in the pedestrian technology of telephony, has meant that the online component of mobile communications has not been experienced as cut off from everyday reality, places, and social identities. (Ito, 2005, p. 8)

This view is consistent with the significance that Howard Rheingold (2002) has attributed to 'smart mobbing', where combining *mobility* with *mobilism* results in the potential for a powerful political activism. The mobility of mobile devices means that they are dispersed throughout the environment, yet their capacity for large-scale *mobilism* is seldom activated, except perhaps at U2 concerts and by advertisers at cultural events (cf. Chesher in this issue). Nevertheless these examples show that the interlocking of mobile communication and the Internet remains important as a counterbalance between the fleeting but highly connected real world of the mobile Internet and the disconnected virtuality of cyberspace.

Mass User Sites

The personal weblog is a one-way communication form by nature of its commitment to individual authorship. It is a web 1.0 form. Another web 1.0 form, the online forum, offers an alternative to the blog model, especially in its capacity to handle mass user memberships. Some Internet forums boast memberships of between two and five million registered users[2] and have attracted hundreds of millions of posts. The online forums also usually host image collection and comment options. The dedicated cameraphone image hosting sites are caught in limbo somewhere between web 1.0 and web 2.0 (O'Reilly, 2005). They have not yet fully embraced social networking, which would require a more effective enactment of *mobilism*, or collectively intelligent activities. They attract significantly fewer registered members than online forums, even though similarly structured. The problem for them has been that their central activity—archiving cameraphone images online—does not generate the same levels of appeal and commitment as forums devoted to, for example, Japanese animation or game cheats. Their primary activity is archiving and cataloguing personal image collections.

We noticed that a few site contributors submitted hundreds and even thousands of images over months and years, but most submitted only a few images and never returned to the site. More accomplished users submitted their images to multiple sites. In some cases, the limitations of the cameraphone seemed to motivate an artistic fascination with the one megapixel camera. We were able to link a few of the more extensive online collections to host site staff. Most site activity was generated by once-only submitters. This suggested that the sites were the first experience of online image archiving for their users, and that the users were exploring what they could do with this new source of image making. The lack of comment on the images suggested that, for most, the initial use of the site expressed an interest in the cameraphone rather than an interest in photography.

Believing that taking better quality photos would encourage people to upload and print more of the images currently stored in phones or downloaded to personal computers, some cell phone manufacturers (e.g. Nokia's Lifeblog) established user sites to encourage users to take better photos, and others (e.g. Motorola) to demonstrate what professional photographers can achieve with a cameraphone (see http://www.sixspace.com/gallery/sent/index.html). While these 'professional' photos demonstrate the limitations of the one megapixel cameraphone, they also show that professional practice is informed by a different rationale and cultural purpose than amateur image-making. Amateur images are inherently documentarist and in this regard incorporate the politics and ideological biases of everyday life. Compared to the professional images at the Motorola site above, amateur images offer a commentary that is missing from the professional images where framing, composition and lighting inform the photographic practice and erase the personal specificity of the images.

Conclusion

I began this research when mobile cameraphone culture was in its infancy and when the Internet still seemed an integral component of mobile culture. When I began writing, mobile culture was recognized as a minority interest on the Internet—one practised primarily in the field (Ito *et al.*, 2005). On the technology front, the distinction between cameraphone and the digital still camera (DSC) is disappearing. I therefore decided to emphasize the network of exchange and the cameraphone's gift economy. The industry gives but it also expects to receive—and the expectation of payback is increasing in ways that may prove too onerous for cameraphone users: surveillance and monitoring; multiple cost structures (access contracts; download limits; site memberships; service costs; file sharing charges). Patronage can overburden creative agency.

Notes

[1] Figure 1 could as easily have been developed using Actor Network Theory, but in this instance I preferred to use concepts of gift exchange and division of labour, which are consistent with the ideas of Gell and Mauss. This has allowed me to emphasize the consensuality in this network, to introduce the concept of patronage, and to identify more clearly the difference between industry and user interests in the cameraphone.

[2] The site www.big-boards.com lists and ranks sites with over 500,000 posts. Gaia Online, an 'anime role playing community' has 5,013,010 registered users and has attracted 835,978,048 posts, while IGN Boards, a general video gaming message board, has 1,255,254 members and has attracted 145,909,298 posts. It is not clear how active these memberships are or over what time period they have accumulated.

References

Akwagyiram, A. (2005) Does 'happy slapping' exist?, *BBC News*, 12 Dec., Available at: http://news.bbc.co.uk/go/pr/fr/-/1/hi/uk/4539913.stm (accessed 27 Mar. 2007).

Born, G. (2006) 'The social and the aesthetic: methodological issues in the study of cultural production', Paper presented at the University of Sydney Department of Media and Communication seminar, Oct.

Fujimoto, K. (2005) 'The third-stage paradigm: territory machines from the girls' pager revolution to mobile aesthetics', in *Personal, Portable, Pedestrian: Mobile Phones in Japanese Life*, eds M. Ito, D. Okabe & M. Matsuda, MIT Press, Cambridge, MA, pp. 77–101.

Gell, A. (1998) *Art and Agency: an Anthropological Theory*, Oxford University Press, Oxford.

Goggin, G. (2005) '"Have fun and change the world": moblogging, mobile phone culture and the Internet', Unpublished paper delivered to Blogtalk Downunder, Sydney, 20–21 May.

Hartley, J. (2005) 'Creative identities', in *Creative Industries*, ed. J. Hartley, Blackwell, Malden, MA, pp. 106–116.

InfoTrends/CAP Ventures (2006) 'Worldwide mobile imaging: stimulating demand and usage', Press Release, Available at: http://www.capv.com/home/Multiclient/mobileimaging2.html (accessed 27 Mar. 2007).

Ito, M. (2005) 'Introduction: personal, portable, pedestrian', in *Personal, Portable, Pedestrian: Mobile Phones in Japanese Life*, eds M. Ito, D. Okabe & M. Matsuda, MIT Press, Cambridge, MA and London.

Ito, M., Okabe, D. & Matsuda, M. (eds) (2005) *Personal, Portable, Pedestrian: Mobile Phones in Japanese Life* MIT Press, Cambridge, MA and London.

Kato, H. (2005) 'Japanese youth and the imagining of keitei', in *Personal, Portable, Pedestrian: Mobile Phones in Japanese Life*, eds M. Ito, D. Okabe & M. Matsuda, MIT Press, Cambridge, MA and London.

Ling, R. (2004) *The Mobile Connection: the Cell Phone's Impact on Society*, Morgan Kaufmann, San Francisco.

Lyra Research Report (2006) 'Dialing for pictures: 2006 U.S. Camera Phone User Survey', Press Release, Aug., Available at: http://lyra.ecnext.com/coms2/summary_0290-461_ITM (accessed 27 Mar. 2007).

M:Metrics. (2006a) 'Captured by cameraphones', Press Release, 7 Aug., Available at: www.webitpr.com/release_detail.asp?ReleaseID = 4328 (accessed 27 Mar. 2007).

M:Metrics. (2006b) 'Teens take user-generated content and social networking to go', Press Release, 14 Dec., Available at: http://www.mmetrics.co.uk/press/Press.aspx (accessed 27 Mar. 2007).

Mauss, M. (1966/1925) *The Gift: the Form and Reason for Exchange in Archaic Societies*, Cohen & West, London.

O'Reilly, T. (2005) 'What is Web 2.0? Design patterns and business models for the next generation of software', Available at: http://www.oreillynet.com/pub/a/oreilly/tim/news/2005/09/30/what-is-web-20.html (accessed 27 Mar. 2007).

Rheingold, H. (2002) *Smart Mobs*, Perseus, Cambridge, MA.

Sinnreich, A. (2007) 'Come together right now: we know something's happening but we don't know what it is', *International Journal of Communication*, vol. 1, pp. 45–47.

Suroweicki, J. (2005) *The Wisdom of Crowds: Why the Many are Smarter than the Few*, Abacus, London.

Zimonjic, P. & Hastings, C. (2005) 'Tortured for £75: a victim of the new form of "happy slapping" designed to instil terror', *The Telegraph*, 29 May, p. 299.

Websites

Andrew Neylan: www.cablog.com.au (accessed 27 Mar. 2007).

Bento: www.flicker.com/photos/mimiito/tags/benrobby (accessed 27 Mar. 2007).

Big Boards: www.big-boards.com (accessed 27 Mar. 2007).

Flickr: flickr.com (accessed 27 Mar. 2007).

Motorola example photos: http://www.sixspace.com/gallery/sent/index.html; view at http://www.sentonline.com/invited-images.html (accessed 27 Mar. 2007).

Reiter's Camera Phone Report: http://www.cameraphonereport.com (accessed 27 Mar. 2007).

TextAmerica: http://www.textamerica.com (accessed 27 Mar. 2007).

The Hand Collector: http://www.robottle.com/hands/ (accessed 27 Mar. 2007).

Mobloguk: http://www.moblog.co.uk (accessed 27 Mar. 2007).

Foneblog: http://www.foneblog.ie (accessed 27 Mar. 2007).

Text-messaging Cultures of College Girls in Hong Kong: SMS as Resources for Achieving Intimacy and Gift-exchange with Multiple Functions

Angel M.Y. Lin & Avin H.M. Tong

Introduction

Mobile text messaging, variously known as SMS (Short Message Service), text messaging, mobile e-mail, or texting, has become a common means of keeping in constant touch, especially among young people in many parts of the world today. The research literature abounds with studies on the social, cultural, linguistic and communicative aspects of mobile text messaging in different sociocultural contexts in the world. However, almost all of these studies have been conducted outside of Hong Kong. Below we shall provide a preliminary literature review of earlier studies related to SMS. Then we shall present our pilot qualitative study. In the findings and discussion section, we shall interpret the findings in light of the relevant research literature.

Linguistic Analysis of SMS Messages

A small but significant set of studies concentrates on the linguistic, semiotic and sociolinguistic analysis of mobile text messages and shows how written language is creatively and effectively adapted to suit the parameters of the new medium for a wide variety of communicative functions. They generally show that contrary to some popular disparaging beliefs that SMS language is 'broken language' and might harm students' literacy skills (e.g. Allison, 2001; Tweedle, 2003; Wilson, 2003), text-messaging language use is actually creative and well suited to the particular mode of SMS communication (e.g. Hard af Sergerstad, 2002). SMS texts are found to attempt maximum economy in script resulting in cryptic, semi-coded messages with hybridized orthography and creative use of various non-letter characters (for example, numerals, emoticons) in spelling/writing (Pietrini, 2001; Galan Rodriguez, 2001). SMS language should therefore be seen as a new, creative, hybridized variety of literacy in its own right rather than a 'broken form' of a traditional print language. Mobile textuality involves a radical fragmentation and unpredictable seriality of text lexia (Goggin, 2004); it is a new form of literacy that is particularly suited to the special features of the new communication medium rather than a degenerated form of print literacy.

Analysis of the Social Impact of New Mobile Communication Technologies

In contrast to the general celebratory optimistic tone of the promotional/advertising discourses of mobile communication companies ('ever closer human relationships', 'closer family ties', 'instant and constant touch'), cultural studies researchers generally tend to stay with a much more cautious and perhaps even pessimistic tone in their theorizing of the social and cultural implications of new mobile communication technologies. For instance, in an essay on 'mobile privatization' Raymond Williams (1983) writes:

> There is then a unique modern condition, which I defined in an earlier book (*Television: Technology and Cultural Form*, 1974) as 'mobile privatization'. It is an ugly phrase for an unprecedented condition. What it means is that at most active social levels people are increasingly living as private small-family units, or, disrupting even that, as private and deliberately self-enclosed individuals, while at the same time there is a quite unprecedented mobility of such restricted privacies. (p. 129)

For Williams the new mobile technologies (mainly referring to the walkman at that time) only serve to further aggravate the modern human condition of 'mobile-privatized social relations'. To illustrate what he means, Williams uses the visual image of car traffic: people insulated in their own moving (mobile) 'shells' (cars) communicate only with their own social networks of acquaintances (people travelling in the private car) in their own mobile 'shells' without any concern for other fellow human beings in other 'shells'. There is an old Chinese saying that captures well Williams' concern about late modernity: 'Each family only sweeps the snow on their own doorway and nobody cares about the frost on other people's roofs.' This seems to be a concern that is aggravated further by new mobile communication technologies, with people talking of 'detached presence'.

This cautious note is echoed in an article by Cooper (2001) on social theory in the wireless world, who concludes that perpetual availability is both an advantage and disadvantage—'You can run but you can't hide.' This cautious note is also found in youth mobile communication studies: young people's constant use of mobile communication (e.g. mobile phoning or text messaging) can be seen as a symptom of a general loss of human connectivity in the modern condition (Lobet-Maris, 2003).

Still others in Finland are concerned about their traditional speech culture being increasingly replaced by a new 'shallow' mobile communication culture (Puro, 2002). Cultural studies researchers might be cautious about Puro's (2002) seeming construction of a static, binary, two-culture theory: old and new cultures each with fixed boundaries. Without falling into the trap of technological determinism, many mobile communication researchers have focused on the complex processes in which mobile communication technologies interact with existing sociocultural practices and on the new technologies as *affording resources* for change in social interactional practices to gradually develop (Katz & Aakhus, 2002). An important force driving these processes might be the human need for management of social roles, social relations, individual and group identities, images and commitments (Goffman, 1959). Recent research studies on mobile communication have examined how SMS can afford a sense of intimacy between friends as well as between strangers.

SMS as a Resource for Intimacy and Shaping Space

The virtual presence (or 'absent presence') of 'persons elsewhere' through mobile communication facilitates networking, deeper relationships, or simply increased contact (Gergen, 2002). People who are physically far away can be brought into immediate cyber presence. This feature of mobile communication is especially appealing to youth, as they consider constant social contact and a sense of co-presence very important. Ling (2004) describes mobile phone as a device providing 'anytime-anywhere-for-whatever-reason type of access' to other members of one's peer group, and this can produce tightly bonded social networks among youths.

In particular, text messaging is more individualized than traditional telephony, as we expect that the message will reach a specific person directly, regardless of where the person is at that moment (Ling, 2004). Ito & Okabe (2005) found that many messages were exchanged for the purpose of maintaining lightweight contact and co-presence rather than exchange of real information—a mode of text sharing they dub 'ambient virtual co-presence'. Hjorth (2005) also examines the generic functions of SMS and its contribution to the perpetual encircling of place, locality and practices of co-presence. It seems that text messaging has created new kinds of modalities for co-presence and communication, which contributes to a sense of virtual intimacy.

The *Longman English Dictionary* (1988) defines the word 'intimacy' as referring to 'the state of being intimate', 'close friendship', 'the act of sex', as well as 'speech or an action such as is permitted only between people who know each other very well'. It thus includes both friendship and romance relationship. As Ito (2002) argues in the context of Tokyo, and Yoon (2003) in the context of Seoul, mobile phone practices are contingent upon the agencies of the local and contribute to maintaining intimacy among friends. On the other hand, some studies also discuss the sense of intimacy between strangers brought about by text exchanges (Ling & Yttri, 2002; Grinter & Eldridge, 2001; Ling, 2000; Prøitz, 2005).

In the Philippines, a Southeast Asian society regarded as the texting capital of the world, text messaging provides young people with an avenue to express their curiosities and desires, especially those about erotic love (Ellwood-Clayton, 2006). Pertierra (2005, 2006) suggests that mobile technologically mediated relationships blur the lines between the real, the surreal and the fantastic, and it opens up possibilities for intimate relationships even between strangers.

Ito & Okabe (2003) studied the kinds of social structural factors that relate to unique patterns of mobile phone usage, particularly text messaging of teenagers in Japan. It was found that Japanese teens' penchant for text messaging is an outcome of a wide range of factors. These include the unique expressive functions and styles of this form of communication but also, most importantly, factors that relate to adults' control and surveillance in particular places. Japanese youth, particularly high school students, move between the places of home, school, and urban space that are all subject to a high degree of regulation and surveillance by adults. Youth peer groups and couples lack ownership and control of place and couples and friends have few opportunities for private conversation.

Mobile text messaging has thus fulfilled an important function which provides a sense of co-presence for young people who lack the means to share some private physical space free from adults' surveillance. Given the relative cultural proximity between Japanese sociocultural norms governing youth behaviour and those of Hong Kong society, where it is equally crowded and difficult for young people to find their own private space free from adult surveillance, it will be interesting to see whether similar or different patterns of youth SMS practices can be found in Hong Kong.

Another study in Taiwan (Yeh, 2004) found that SMS is used to negotiate subtle gender relations, especially among couples (e.g. after quarrelling, the other does not want to hear one's voice; avoiding embarrassment when making romantic advances or when saying 'no' to such advances). The informants also expressed the fact that some messages are highly private and some are highly meaningful, which they keep.

Thus, both studies on East Asian youth text messaging end on an optimistic note about the positive uses of SMS by young people for gaining freedom from surveillance by adults or for negotiating subtle gender relations. This is in line with Goggin's (2004) observation that young people took to text messaging as a tactic of consolidating their shared culture, in distinction from the general culture dominated by their parents and other adults.

Another important feature of SMS exchange practice is that it somewhat resembles the 'gift-exchange' practice in the offline world (Johnsen, 2000; Pertierra, 2005, 2006). Social life in traditional society involves lots of exchanges, which concern both material and symbolic goods (Mauss, 1969). Mauss argues that the act of gift giving is essentially the recognition of the moral basis of social life, which recognizes that people are entailed in one another in a relationship of mutual ontology.

As Taylor & Harper (2003) point out in their study of youth usage of mobile phones, the practice of text messaging creates a level of reciprocity best understood as an extension of the gift-giving practice. Pertierra (2005) suggests that in text messaging someone initiates the exchange and expects the other side to reciprocate: 'The text as gift is an important aspect of an economy of intimacy and desire' (2005, p. 7). Hjorth (2005) uses the metaphor of 'postcard' to describe the symbolic function of text messaging, as both of them include gestures of intimacy and hallmark clichés, and involve unspoken forms of obligation and reciprocity.

The Increasing Popularity of Text Messaging (SMS) in Hong Kong

For many years Hong Kong has been one of the places with the highest penetration of mobile phone service in the world. From 1998 to 2003 the number of mobile service subscribers had increased 1.5-fold. The number reached 7.19 million by the end of 2003, representing a penetration rate of 106 per cent (source: www.info.gov.hk). Despite this high mobile phone penetration, SMS is not as widespread as in other economically developed Asian societies such as Singapore, the Philippines or South Korea. The TNS Asia Telecoms Index shows that only 43 per cent of Hong Kong cellphone users use SMS and the average number of messages sent per user per month is only 23. This is perhaps due to the fact that Hong Kong was relatively late in introducing inter-operator SMS. There has also been little promotion of SMS by the service providers in Hong Kong. Another possible reason is that mobile phone calling is relatively cheap in Hong Kong compared with other East Asian cities and so people do not need to use SMS to save on phone bills. One sociocultural reason might also be the fact that, unlike other East Asian societies such as Japan (Ito & Okabe, 2003) and Korea (Kim, 2002), talking loudly (especially by adults) over the mobile phone in Hong Kong public areas seems to be a common habit among many Hong Kong people and there does not seem to be great sociocultural pressure on Hong Kongers to switch their loud mobile phone talking mode to SMS mode so as not to disturb other people in public places such as the subway or the bus. SMS can thus be said to be a still largely under-developed area in Hong Kong.

However, Hong Kong youths are catching up fast in their SMS consumption. According to the statistics compiled by the Office of the Telecommunications Authority of Hong Kong, 116 million SMS messages were sent in January 2005, as compared to 23 million in December 2003, representing a sharp increase of 388 per cent in two years (*Hong Kong Economics Journal*, 2005).

However, there has been only one study on youth SMS practices in Hong Kong. Lin & Lo (2004) conducted a pilot study on the SMS practices of Hong Kong college students, and they found significant gendered differences in using SMS (e.g. significantly more females using SMS than males). Dating and social grooming with boyfriends/girlfriends also seem to be chief reasons for young people's use of SMS. As it was a quantitative study, it is difficult to discern the deep social meanings of SMS practices. Thus it is in the spirit of preliminary, situated exploration that a pilot qualitative study on the SMS practices of Hong Kong college students was conducted between September and October 2005. The study aimed at taking an initial look at what roles SMS may play in the everyday life of some young educated college students (particularly girls) in Hong Kong.

The data collection methods consisted of individual in-depth interviews and end-of-interview questionnaires. The interviews were conducted in the period September to October 2005, with twelve female informants and three male informants (two of them being partners of two of the female informants). All quotes used in the paper were originally in Cantonese and were translated into English. As a pilot study, the small sample poses a major limitation, but it does provide us with some initial insights on the emerging SMS practices among young college women in Hong Kong.

SMS as a Resource in Achieving a Sense of Co-presence and Intimacy by Young College Students

One of the conditions for maintaining a sense of co-presence between friends is that of being easily available to each other. Our informants were asked to compare SMS with other communicative devices, and most of them said that SMS has some unique advantages which enable users to be more 'readily available' to each other. For instance, Katie found that SMS makes her easily accessible to her friends in all situations. 'Except when you have a computer in front of you at home, otherwise I will use SMS. Because it is convenient!' Another informant, Mandy, explained: 'I think SMS can make sure that he/she can receive (the message) immediately . . . It's different from MSN and ICQ, with which you can read the messages only when you are online.' This reveals one of the strengths of SMS—convenience and easy access most of the time. While this type of perceptual contact is pervasive, it is less disturbing than other more direct means of communication such as phone calls. The use of SMS successfully avoids such kind of disturbance, as it is compatible with the multi-tasking practices of youths in their everyday lives.

In addition to its easy availability and non-disturbing nature, our informants found that SMS affords a sense of closeness and intimacy. For instance, Bobo pointed out that SMS is more 'cordial and intimate' than MSN, 'Maybe SMS gives you a feeling that it is on the go. There are no constraints; you don't need to sit in front of the computer . . . and you can even do SMS messages in the washroom.' This provides them with a kind of virtual co-presence which forms a strong binding force between the sender and the receiver.

Friends are crucial to young people in contemporary society, and the use of SMS ensures their availability and accessibility. In our study, SMS was used by the informants to arrange social gatherings and to send greetings to friends during festivals. It shows that text exchange enables young people to sustain and strengthen existing relationships. Many informants regarded it as the best tool for keeping close and intimate relationships with their old friends whom they do not see too often due to physical distance (e.g. former classmates in secondary schools).

SMS communicative practices thus provide them with a sense of intimacy and togetherness with their old friends, whom they cannot see regularly.

In addition to maintaining existing social relations, SMS also provides a channel for users to develop new friendships with newly acquainted people, allowing them to get to know their new friends gradually in the textual world. Ling & Yttri (2002) suggest that SMS helps one to explore potential new friends, and the friendship is situated in a more careful and controllable context. Sung & Yeung (2005) also found that Mainland Chinese college students make frequent phone calls only to classmates and lovers, but in the case of SMS they have a wider range of targets. Many of our informants agreed that SMS facilitates more open communication and increases their opportunities for contacting new, unfamiliar friends, including those who may be further developed into romantic partners. It encourages new contacts and enlarges one's social network by providing a form of pervasive but non-disturbing contact. Most important of all, it is particularly useful for girls to employ tactics in initiating a new, intimate relationship.

The exchange of text resembles online chat in turn taking and discourse structure (Kasesniemi & Rautianien, 2002), which is different from direct face-to-face or voice-to-voice interaction. It affords a form of intimacy but also a 'safe' distance between the users. On top of a sense of co-presence, it offers both time and space for buffering. It is believed that asynchronous text-based communication allows users time to select, craft out, and edit the personality they want to present in their messages. In our study, some informants appreciated the asynchronous text-based nature of SMS, which can be utilized as a powerful tool for girls. For instance, Linda explained this with an example of approaching a stranger: 'If you have just known someone, you do not know what to say to him on phone. If you send him a text message, you can just type one sentence ...'

In fact, SMS can help one to initiate ambivalent relationships, and some of them can be further developed into offline romance relationships. Ellwood-Clayton (2003) found that some Filipinos are restructuring amorous lived relationships and creating virtual ones through text communication. In this study, SMS seemed to provide a space for further developing emerging amorous feelings and romance, and it was found to instil personal bravery among its users. One of the female informants, Suki, believed that SMS marks the starting point of a close relationship, as it allows her to understand this male friend more deeply in a non-embarrassing context. She illustrated it with her own experience: 'It really made us closer! It is because he is not so talkative, but when he sends me SMS, (I discover that) though he usually speaks very little, he can actually express a lot in SMS.'

In particular, women are encouraged to express their feelings and take more initiative in the textual world, which seems to subvert the traditional gender roles in courtship among young Chinese. They tend to develop intimate relationship more easily through this pervasive but indirect contact. Some of our female informants found that they can be more active and expressive when they chat with male friends in SMS. As a result, they consider it to be the 'best communicative means' in starting a relationship.

Due to the absence of visual and vocal signals, girls are encouraged to express their heartfelt feelings more bravely, particularly in the initial stages of courtship. Another female informant, Nancy, described SMS as a mixed form of 'passive' and 'active' communication styles: 'You can ignore his reaction (after receiving your message) ... no need to care whether he replies to you or not, but you can let him know what you think.' This shows that SMS can be used as a 'tactic', as a kind of 'strategy', especially by Chinese girls, to develop a closer relationship with a male friend. Though the relationship may be confined to the textual world at the beginning, many of them do extend it to the offline world after a period of time. This reveals that SMS enables them to negotiate relationships with males in an 'intimate' but also 'distanced' and safe space.

SMS as an Ideal Space for Communicating with Lovers

Many studies have shown that the cell phone provides users with a site to explore their desires, as compared to traditional face-to-face communication (Ellwood-Clayton, 2003; Kasesniemi & Rautianen, 2002; Pertierra *et al.*, 2002). In our study, it was found that SMS provides a site for sustaining two types of romantic relationship. The first type is the ambivalent relationship that is still emerging in the early or 'underground' stage. Another type is that of romantic partners. As revealed by some of our informants, SMS allows new forms of intimacy between lovers. It provides a platform for transmission of both 'unsayable' and ordinary topics, and also the exchange of various 'poetic' quotes that people seldom mention in other situations. As suggested by Pertierra (2006), the requirement for privacy is crucial for many aspects of modernity. Since mobile phones are very personal possessions, text exchange thus provides a sense of privacy for the sender and receiver, and it allows them to express their inner feelings in this private space more easily.

Some informants considered SMS to be particularly helpful for Chinese people, since it is said that the Chinese are less willing to express their inner worlds directly by speech. For instance, Nancy said: 'I think SMS allows you to write from your heart, not speak from your heart ... Sometimes may be Chinese, or Eastern people are like this. I find something difficult to speak out, cannot be spoken, then (I will) use, I used letters before, and now use these three tools (SMS, MSN and ICQ).' Yuko believed that Chinese people need some channels to reveal their inner world, and SMS provides them with this opportunity: 'Chinese are relatively more reserved ... those real feelings in your heart, (you will wait till) a particular situation, a particular moment, then you will finally speak it out. But these situations may not be there all the time, so you can make use of text-messaging.'

Haig (2002) suggests that SMS provides privacy, as the conversations remain private even in public locations, which makes it totally different from making direct phone calls. As a result, people may tend to reveal more about their emotional selves in the textual space. One of the illustrative examples of the private, 'unspoken' topics is sweet words and honeyed phrases among lovers. An informant, Bobo, remarked that she will say something 'very sweet' in SMS messages. She used to call her partner 'xiang gong' (a formal Chinese term to address one's husband in ancient times; e.g. as found in Chinese historical dramas), and her boyfriend used to call her 'niang zi' (a formal Chinese term to address one's wife in ancient times; e.g. as found in Chinese historical dramas).

It is interesting to note that both females and males in our study found it easier to 'speak out' their feelings and exchange sweet words with their partners through text messaging. This is consistent with Ellwood-Clayton's (2006) finding that mobile phones provide young people with an avenue to express their desires, and to deepen intimate relationships.

Symbolic and Emotional Functions of Text as a Gift

According to our informants, around two-third of the messages are used for the maintenance of friendship, namely romantic and social functions associated with intimate and relational concerns, and only one-third of them are used for practical information exchange. It seems that text messaging plays an important role in enabling its users to share feelings of intimacy in a gift-exchange process.

In our study, all of our female informants participated in the practice of sending season's greetings and birthday greetings to their friends, partners, and family members. It seems that such greetings fulfil the care-taker and expressive roles of girls expected in some cultures. In some cultures, the everyday tasks of mothering, such as the routine care of home and children, are taken to be the natural expressions of femininity (Coltrane, 2004); expressing tender feelings and being gentle are also considered to be ideal qualities for women (Cancian, 2004). These kinds of greetings resemble the role of gifting during special seasons and festivals. In expressive messages, texting can be a form of intimate sharing between users.

It seems that texting is now a key tool in sustaining the emotional lives of young people (Vincent, 2003), and many informants consider SMS as an essential part of their social life. Pertierra (2006) considers this expression of gift exchange as laying the basis for friendship; and in the practice of text messaging, receiving messages is a confirmation of one's membership in the group (Ling & Yttri, 2002). As illustrated by some of our informants, it acts as an indicator of close relationship or one's ability to be likeable and liked. In fact, this confirmation of close relationship requires some effort in crafting a message. The kind of 'gift' exchanged should symbolize the specific relationship between the sender and the receiver to make the relationship more private and intimate. It should be noted that many informants emphasize originality and creativity in composing text messages (particularly birthday and season's greetings). The act of sending 'personal' messages (at least perceived to be personal) possesses a kind of symbolic significance in confirming one's important position, and the frequency of text exchange is highly related to the popularity of a person in a social circle, as perceived by our young informants.

Apart from the function of establishing reciprocal relationships, one important feature of a gift is that it can be kept for future remembrance. In this study, all of our informants reported the practice of saving messages (particularly sweet words from partners, encouragement from good friends, and also season's greetings) in their original form and preserving the embodied memories. This is regarded as a common practice in youth social networks. As indicated by some of our informants, these expressive messages perform a very crucial emotional function not only at the time of exchanging the text but also at the time they re-read the messages when in need.

Conclusion

Text messaging seems to open up the possibility of openness, as it encourages new contacts and enlarges one's social networks. It is similar to the idea that SMS resembles a virtual cafeteria or marketplace where people can meet each other (Puro, 2002; Vincent, 2003). And this virtual cafeteria is carried in one's pocket and can be accessed anytime, anywhere. As shown in our findings, young Hong Kong Chinese college girls find SMS very useful in gaining friends by encouraging them to take the initiative to start new relationships while sustaining previously established friendships. SMS might thus serve as a catalyst in helping men and women to communicate more uninhibitedly and to develop and adopt converging communicative styles (that is, crossing traditional gender role boundaries).

In short, SMS helps maintain social relations in modern society, and affords resources to achieve a sense of co-presence and intimacy with both existing friends and new acquaintances, while avoiding having to deal with the fluster and embarrassment of a face-to-face interaction or the intrusive disturbance of a phone call. In our study, the young women expressed their preference for using SMS to express their caring feelings and to send greetings to others (including partners and friends). They also found it less disturbing and more considerate to contact others through text messaging. This fits the image of the care-taker, a role that is often expected to be performed by a female in some cultures (and the socialization experiences of the authors in Chinese society support this observation). In this respect, the new mobile communication technology seems to have inserted itself into existing sociality and discursive practices without initiating any radical transformation of young women's ways of life.

It seems that SMS has not radically changed the existing gendered pattern of sociality practices, at least from the point of view of the preliminary data of our pilot study. However, negotiation of gender roles and practices that are slightly different from traditional norms seems to be actually taking place at the same time. Modern women are now exploring new roles and identities made viable by SMS, where they can shift and play with new roles, being more active and expressive, and even subverting traditional gender roles by initiating a romantic relation. On the other hand, it seems that SMS can also bring about change and breakthroughs for young men, as our male informants (though limited in the sample size) reported exchanging sweet words and inner feelings with friends (particularly girlfriends and friends of the opposite sex) more readily in the SMS textual world.

To conclude, it seems, then, that this new mobile communication technology is useful for modern young women to sustain different types of social relationships in a modern society that experiences frequent relocations of people, and it may also interact with existing sociocultural and discursive practices to produce gradual change in the traditional gendered patterns of communicative practices. Moreover, it might be that this new mobile communication technology can increase the agency of young modern Chinese women, at least in encouraging them to take more initiative in self-expression and relationship building (particularly in starting and sustaining romantic relationships). This transformation of ways of being may be gradual and may emerge invisibly as more and more young people (including males) become SMS users for a broader ranger of purposes. Further research studies should investigate both female and male informants through the use of ethnographic studies, communication diaries, and focus group interviews, in order to study the different cultural and social practices of male and female users.

References

Allison, R. (2001) 'Phone text chat "harms literacy"', *Guardian*, Jan., 22, p. 7.

Cancian, F. M. (2004) 'The feminization of love', in *The Gendered Society Reader*, eds M. S. Kimmel & A. Aronson, 2nd edn, Oxford University Press, New York and Oxford.

Coltrane, S. (2004) 'Household labor and the routine production of gender', in *The Gendered Society Reader*, eds M. S. Kimmel & A. Aronson, 2nd edn, Oxford University Press, New York and Oxford.

Cooper, G. (2001) 'The mutable mobile: social theory in the wireless world', in *Wireless World: Social Aspects and Interactional Aspects of the Mobile Age*, eds B. Brown, N. Green & R. Harper, Springer, London.

Ellwood-Clayton, B. (2003) 'Virtual strangers: young love and texting in the Filipino archipelago of cyberspace', in *Communications in the 21st Century. Mobile Democracy: Essays on Society, Self and Politics*, ed. K. Nyíri, Passagen, Vienna.

Ellwood-Clayton, B. (2006) 'All we need is love—and a mobile phone: texting in the Philippines', in Paper presented at the *International Conference on Cultural Space and the Public Sphere in Asia*, Seoul, South Korea, 15–16, Mar.

Galan Rodriguez, C. (2001) 'SMS? A new language model?', *Español Actual*, vol. 76, pp. 93–103.

Gergen, K. J. (2002) 'The challenge of absent presence', in *Perpetual Contact: Mobile Communication, Private Talk, Public Performance*, eds J. E. Katz & M. Aakhus, Cambridge University Press, Cambridge.

Goffman, E. (1959) *The Presentation of Self in Everyday Life*, Doubleday, Garden City, NY.

Goggin, G. (2004) 'Mobile text', *M/C: A Journal of Media and Culture*, vol. 7, Available at: http://www.media-culture.org.au/0401/03-goggin.html (accessed 28 Nov. 2006).

Grinter, R. & Eldridge, M. (2001) 'Y do tngrs luv 2 txt msg?', in *Proceedings of the Seventh European Conference on Computer-supported Cooperative Work ECSCW 01*, eds. W. Prinz et al., Kluwer, Dordrecht.

Haig, M. (2002) *Mobile Marketing: the Message Revolution: 'A Cracking Insight into How to Really Get Personal with your Customers*, Kogan Page, New York.

Hard af Sergerstad, Y. (2002) *Use and Adaptation of Written Language to the Conditions of Computer-mediated Communication, Unpublished doctoral dissertation*, University of Goteborg.

Hjorth, L. (2005) 'Locating mobility: practices of co-presence and the persistence of the postal metaphor in SMS/MMS mobile phone customization in Melbourne', *Fibreculture Journal*, no. 6 Available at: http://journal.fibreculture.org/issue6 (accessed 28 Nov. 2006).

Hong Kong Economics Journal (2005) 'A flood of SMS', 5 Apr., p. A15.

Ito, M. (2002) 'Mobiles and the appropriation of place', *Receiver Magazine*, no. 8, Available at: http://www.receiver.vodafone.com/ (accessed 28 Nov. 2006).

Ito, M. & Okabe, D. B. (2005) 'Technosocial situations', in *Personal, Portable, Pedestrian: Mobile Phones in Japanese Life*, eds M. Ito, D. Okabe & M. Matsuda, MIT Press, Cambridge, MA.

Ito, M. & Okabe, D. (2003) 'Mobile phones, Japanese youth, and the re-placement of social contact', Paper presented at the *Front Stage–Back Stage: Mobile Communication and the Renegotiation of the Social Sphere* conference, Grimstad, Norway, 22–24 Jun.

Johnsen, T. E. (2000) *Ring Meg! En Studie Av Ungdom Og Mobiltelefoni*, Department of Sociology, University of Oslo.

Kasesniemi, E. L. & Rautianen, P. (2002) 'Mobile culture of children and teenagers in Finland', in *Perpetual Contact: Mobile Communication, Private Talk, Public Performance*, eds J. E. Katz & M. Aakhus, Cambridge University Press, Cambridge.

Katz, J. E. & Aakhus, M. A. (2002) 'Conclusion: making meaning of mobiles—a theory of Apparatgeist', in *Perpetual Contact: Mobile Communication, Private Talk, Public Performance*, eds J. E. Katz & M. Aakhus, Cambridge University Press, Cambridge.

Kim, S. D. (2002) 'Korea: personal meanings', in *Perpetual Contact: Mobile Communication, Private Talk, Public Performance*, eds J. E. Katz & M. Aakhus, Cambridge University Press, Cambridge, pp. 63–79.

Lin, A. & Lo, J. (2004) 'New youth digital literacies and mobile connectivity: text-messaging among Hong Kong college students', Paper presented at the *International Conference on Mobile Communication and Social Change*, Seoul, South Korea, 18–19 Oct.

Ling, R. (2000) 'Direct and mediated interaction in the maintenance of social relationships', in *Home Informatics and Telematics: Information, Technology and Society*, eds A. Sloane & F. V. Rijn, Kluwer, Boston.

Ling, R. (2004) *The Mobile Connection: the Cell Phone's Impact on Society*, Morgan Kaufmann, San Francisco.

Ling, R. & Yttri, B. (2002) 'Hyper-coordination via mobile phones in Norway', in *Perpetual Contact: Mobile Communication, Private Talk, Public Performance*, eds J. E. Katz & M. Aakhus, Cambridge University Press, Cambridge.

Lobet-Maris, C. (2003) 'Mobile phone tribes: youth and social identity', in *Perpetual Contact: Mobile Communication, Private Talk, Public Performance*, eds J. E. Katz & M. Aakhus, Cambridge University Press, Cambridge.

Longman English Dictionary (1988) Longman, Harlow.

Mauss, M. (1969) *The Gift*, trans. I. Cunnison, Cohen & West, London.

Pertierra, R. (2005) 'The gift of the text', in Paper presented at the *Asian Digital Cultures Conference*, Taipei, Institute of Ethnology, Academia Sinica, 28–29 Jul.

Pertierra, R. (2006) 'The transformative capacities of mobile phones and the Internet—a Philippine example', *Paper presented at the International Conference on Cultural Space and the Public Sphere in Asia*, Seoul, South Korea, 15–16 Mar.

Pertierra, R. *et al.* (2002) *Txt-ing-Selves: Mobile Phones and Philippine Modernity*, De La Salle University Press, Manila, Available at: http://www.finlandembassy.ph/texting1.htm (accessed 28 Nov. 2006).

Pietrini, D. (2001) 'The SMS and the triumph of informality and ludic writing', *Italienisch*, Vol. 46, pp. 92–101.

Prøitz, L. (2005) 'Cute boys or game boys? The embodiment of femininity and masculinity in young Norwegians' text message love-projects, *Fibreculture Journal*, no. 6, Available at: http://journal.fibreculture.org/issue6 (accessed 28 Nov. 2006).

Puro, F. P. (2002) 'Finland: a mobile culture', in *Perpetual Contact: Mobile Communication, Private Talk, Public Performance*, eds J. E. Katz & M. Aakhus, Cambridge University Press, Cambridge.

Sung, S. & Yeung, S. W. (2005) 'Self orientation of college students and consumption of cell phones', *Journal of Social Theory*, vol. 8, no. 1, pp. 3–19.

Taylor, A. & Harper, R. (2003) "Age-old practices in the "new World": a study of gift-giving between teenage mobile phone users", *Changing Our World, Changing Ourselves: Proceedings of the SIGCHI Conference on Human Factors in Computing Systems*, Minneapolis, ACM Press, New York, pp. 439–446.

Tweedle, M. (2003) 'Correct spelling', *The Times*, 1 Oct., p. 17.

Vincent, J. (2003) 'Emotion and Mobile Phones', in *Communications in the 21st Century. Mobile Democracy: Essays on Society, Self and Politics*, ed. K. Nyíri, Passagen, Vienna.

Williams, R. (1983) 'Mobile privatization', Reprinted in P. Gay, S. Hall, L. Janes, H. Mackay & K. Negus (1997) *Doing Cultural Studies: the Story of the Sony Walkman, Sage, London*, p. 1997.

Wilson, D. (2003) 'Texting threatens the correct use of English', *South China Morning Post*, 23 Sept., p. 5.

Yeh, M. -J. (2004) 'A preliminary study on SMS use of youth tribes', *Information Society Research*, vol. 6, pp. 235–282, (original in Chinese).

Yoon, K. (2003) 'Retraditionalizing the mobile: young people's sociality and mobile phone use in Seoul, South Korea', *European Journal of Cultural Studies*, vol. 6, no. 3, pp. 327–343.

Mobiles into Media: Premium Rate SMS and the Adaptation of Television to Interactive Communication Cultures

Christina Spurgeon & Gerard Goggin

Introduction

In September 2005, Li Yuchun, a 21-year-old music student from Sichuan province in southern China, rocketed to fame as winner of Hunan Satellite TV's *Super Girl* (*chaoji nüsheng*), an *Idol*-like talent quest. Incredibly, 3.5 million people, including many expatriate Chinese, voted for her, most using SMS (Short Message Service). Li's popularity, despite her non-conventional appearance and off-key singing, demonstrated the power of SMS to mobilize a massive support base. In all, 8 million people used SMS to vote for 1 of 15 finalists (Jakes, 2005).

In the same year similar extraordinary effects were achieved on a global scale during the Live8 concerts. An estimated 30 million people in 130 countries used the Web, e-mail and SMS to respond to a global call to action to alleviate poverty in the poorest countries of the world.[1] Another example of transcontinental cross-media interaction is the *Eurovision Song Contest*. Since 2004, popular votes have been cast exclusively by fixed or mobile telephone.

The use of SMS as a feedback channel or return path (that is, a new communication channel) in broadcasting is increasingly commonplace. But what is the real significance of these SMS-mediated television events? Is this massive shift in the new digital mobile messaging landscape a lifeline for broadcast television, now struggling with its advertising base? How are new developments in SMS and MMS (Multimedia Messaging Services) building synergies between old and new media? How do they extend mobile phone culture? We address these questions by identifying factors that are shaping the rapid development of SMS as a powerful technology of cooperation (Rheingold, 2002). We also consider how these ubiquitous technologies are implicated in the development of new circuits of culture and new media forms (Marshall, 2004), specifically SMS-enabled interactive television formats.

Tensions within globalizing and converging media services and markets are shaping new cross-media applications. On the supply side, communications carriers are looking to develop new consumer markets for high-value data services, and broadcasters are looking to diversify revenue streams. On the demand side, there is strong consumer interest in opportunities to shape and influence, as well as make, media content. For its part, SMS harnesses consumer-citizen productivity in content and revenue-generation applications at the intersection of broadcast media and mobile communications services.

Convergence in media and communications services requires a more dynamic theorization of subjectivity. Although this theme is not our key focus it is nonetheless necessary to locate this discussion in relation to some of the key developments in critical scholarship of new media subjectivity. Both the solitary broadcast audience member and the mobile communications consumer no longer make sense. Terminology is struggling to explain the new media environment. New propositions abound, including the 'consumer-citizen' of DIY media cultures (Hartley, 1999); the 'produser' of online social networks such as blogs (Bruns & Jacobs, 2006); the 'prosumer' of participatory fan cultures (Toffler, 1970; Jenkins, 2006; Marshall, 2004, p. 25); and the 'viewser' of computer games (O'Regan & Goldsmith, 2002, p. 103). Amongst other things, these awkward neologisms indicate a narrowing of the distance between service producers and consumers; furthermore, they are characteristic of globalizing and converging media and communication environments (Marshall, 2004, p. 25).

The integration of SMS into broadcast television formats narrows the gap between producers and audiences/consumers/citizens. Conversely, the value chains associated with such cross-media SMS applications are often more complex than other new media and communications services. To apply terms developed to describe the changed relations of production in social networks such as blogs, fan cultures or computer games, to SMS-enabled interactivity in television formats is to risk overstating individual consumer agency in the commercial cross-media context. The SMS activity in the latter is more about the liberation of the passive couch potato: transformed into voters—citizenship with a difference.

Another important point of difference is that the commercialization of social networks such as blogs, fan cultures, and computer games tends to proceed on a niche market basis, albeit often on a global scale. While the historical rationale of the *Eurovision Song Contest* was to provide a space for pan-European experimentation in wide area television interactivity (Wikipedia, 2006a), more recent formats, such as *Big Brother*, were explicitly commercial from the outset. Real-time global mass market reach is logistically possible as the case of Live8 demonstrates—although for the time being a successful commercial application of this capacity remains elusive. It remains the case that in the process of achieving mass appeal, audiences for these new media services are transformed into users (Marshall, 2004, pp. 44–51).

Given the primary interest here in the cultural consequences of mobile interactivity rather than questions of cultural citizenship, we initially settled on the term 'end-user' to mark the consequences for subjectivity that arise in the changed relations of production and consumption. We have been subsequently persuaded that this term is inadequate for this task because it 'strips the other composite descriptors of their cultural specificity—as viewers, artists or whatever'.[2] We have elected instead to use the term 'consumer-citizen' because it permits this kind of multi-dimensionality that industry-centred discourses about 'end-users' actively constrain. Indeed, we have retained the term 'end-user' in places to allow precisely this contrast. The term 'citizen-consumer' also points to the larger discussion within cultural studies about the conflation of consumption and citizenship practices that reality television formats such as *Big Brother* demonstrate (e.g. Nightingale & Dwyer, 2006). Although we do discuss *Big Brother* in some detail, we pursue a different thread of analysis and argument.

Broadcast television markets have matured in many countries with some suggesting that television is in decline, and innovation being seen as one of the answers. Is the mobile phone *the* enabling technology for cross-media innovation? Certainly, its uptake suggests it may be the communications technology of the future. A much younger technology than broadcast television, the mobile phone has achieved rapid, though uneven, worldwide adoption and diffusion. Meanwhile, markets for mobile carriage services and handsets have matured with surprising speed. The ways in which mobile communications are being used—and the cultural transformations in which they are implicated—are rich areas for media, communication, and cultural studies research.

On the supply side, competitive market environments are powerful *drivers* in the development of mobile communications. Private investment in infrastructure and spectrum has been responsible for rapid expansion of services. *Inhibitors* include cooperation—or more precisely lack of cooperation in facilitating network interconnection and transparency. Although demand-side influences are contingent on the social and cultural contexts of consumer-citizens, network effects also explain the exponential growth in mobile connectivity (Economides, 1996; Reed, 1999). In other words, increased usage leads to increased value for everyone in the network. This is both tangible and intangible; for instance lower usage costs and greater access to a variety of content. Ultimately, aggressive supply-side marketing of handsets and services to targeted demographics is only likely to *influence* a consumer's brand choice. What is decisive in the consumer-citizen's appropriation and domestication of the technology, and the shaping of mobile phone culture associated with it, is enhanced control, interactive communication, and the habitus of anywhere, anytime connectivity (Middleton, 2002).

In the first part of the paper the focus is on mature markets where competition for mobile services has been so intense that both handsets and services have been quickly reduced to commodity status. We consider the positioning of data services in the marketing strategies of carriers and handset manufacturers, with remarks on China, the largest mobile phone market (in terms of volume of handsets) in the world. We pay particular attention to data services charged at premium rates, which involve a number of telecommunications, multimedia, Internet, television, and content providers in their 'value chains'. In the second part of the paper we consider why this was so, and why carriers have persisted with these strategies in the face of apparent consumer-citizen resistance to many of these services. The third section of the paper examines the successful incorporation of telephone-based interactivity into TV programme formats. We conclude with some remarks on how these developments in modes of mobile media demonstrate consumer-citizen interest in the capacity for a greater variety of modes of interaction with media.

Avoiding the Commodity Trap: the Origins of Mobile Data Services

The mobile phone is one of the fastest growing communication technologies ever. A consequence of rapid growth is that, once established, many mobile markets quickly reach maturity (Spurgeon, 2005). Cut-throat competition for the highly substitutable commodities of handsets and bandwidth means that carriers and manufacturers can try to compete on price alone ('the commodity trap') unless they differentiate their offerings from those of competitors. Competing on price frequently leads to declining margins and diminishing rates of return on investment. Branding is an important alternative marketing strategy that seeks to check the downward spiral associated with competing on price. It seeks to differentiate a product by inscribing it with values and attributes that will appeal to specified consumer groups (target markets). These values and attributes may, or may not, relate to price.

In many countries, carriers and manufacturers turned first to the lifestyle attributes of mobile connectivity and handset design to avoid the commodity trap. They leveraged brand values and market position—a classic example was the iconic Nokia design during the consolidation of second-generation (2G) digital mobile telecommunications in the 1990s (Cunningham, 2005; Goggin, 2006; Häikö, 2002; Lindholm *et al.*, 2003). Emphasis on design and lifestyle attributes of mobile handsets and connectivity alone no longer provides the leverage it once did in crowded markets. Global carriers and manufacturers have initiated a buffet of customizable value-added data services as part of their evolving branding strategies (Wang, 2005; Wilson, 2006). Ringtones and wallpapers provided ways for consumer-citizens to customize their handsets. Games allowed interaction with multiple players. Cameras supported instant communication and exchange of audiovisual formats. Importantly, from the carrier perspective, many of these services and applications also held the promise of stimulating consumer interest in making more intensive uses of expanding amounts of bandwidth. Manufacturers also anticipated demand for new handset designs, which incorporated increased memory and functionality.

The rapid evolution of the mobile telephone from a business communications platform to consumer commodity, to individual lifestyle statement, to mobile networked information and entertainment device, is nowhere more pronounced than in the People's Republic of China. Large numbers of Chinese consumers have bypassed fixed-line service connection costs and queues to leapfrog into the globalizing networked society. In a very short period of time China has emerged as the world's largest consumer market for mobile handsets and services, with more than 350 million mobile subscribers. China is now the global centre of mobile phone manufacturing. Many features of the Chinese mobile phone market are common the world over, including the dominance of global brands such as Motorola (USA), Nokia (Finland), Siemens (Germany) and Samsung (Korea). There are also distinctive local features, which, due to the sheer size and complexity of the Chinese market, have global impacts.

One of these features is the large number of local Chinese handset manufacturers. Over 40 local and international brands tout hundreds of handset models (ABC, 2004). Consequently, competition for market share often occurs on the basis of price rather than brand values (Keane & Spurgeon, 2005). In this highly competitive, maturing market, where competition on price has been particularly intense, transnational handset manufacturers and carriers quickly turned to new content and applications services to maintain a competitive advantage. According to Chinese advertising industry scholar Jing Wang, music content on mobile devices became a new strategy based on the premise that 'musical taste' was an important demographic index for youth marketing in developed countries (2005, p. 188). Motorola was the first international brand to launch such a strategy in China. In 2002, it rolled out its Moto campaign, aiming to establish the brand as the mobile entertainment choice for 'cool' Chinese youth. Motorola entered into joint ventures with MTV and Apple iTunes to enhance its music offerings. It also developed the 'Emerging Artists' platform for the purpose of discovering local new music talent and thereby cement relationships with an elusive but highly desirable youth market. This initiative was not an instant success for a variety of reasons; assumptions about the global similarities of youth on the part of Motorola did not translate to China; insufficient attention had been given to researching Chinese middle-class youth. Wang's own research found that 'age' was a far more reliable index than 'musical taste' for brand choices made by this group. Other factors, explored in the next section, may also explain apparent consumer-citizen resistance to these premium rate mobile data services.

The Limits of Consumer-led Innovation in a Premium Rate Culture

The increase in bandwidth that has accompanied the proliferation of new media technologies, particularly over the past decade, has fed anxieties about how this capacity is most likely to be used. A number of myths have gained currency in many corporate strategies and public policy contexts, including the proposition that 'content is king' in new media environments. In fact, there is a growing body of literature that identifies connectivity as a major driver in the adoption and diffusion of new media technologies (e.g. Reed, 1999; Middleton, 2002), including mobile telephony (Wilson, 2006; Odlyzco, 2004).

In the 'always-on' environments of third-generation (3G) mobile networks, mobile data services have emerged as a key strategy for developing and maintaining high rates of return on investment. With value adding to mobile communications the means to competitive advantage, new intermediaries have moved into position; there are high hopes for the success of location-based applications and the innovations that wireless broadband capabilities will support. Having paid high prices for the 3G spectrum, especially in European markets, mobile carriers have committed to strategies that rely upon consumer demand for streamed content services, and a willingness to pay a premium for these services. What have been neglected are potential markets in peer-to-peer communication of content, and the possibility of consumer-generated media being created and put into circulation via mobile data networks (Wilson, 2006). A shared industry perspective is that mobile data services development will be supply-side driven and will occur principally in carrier-controlled proprietary spaces. Such a stance largely neglects the experiences of the Internet where innovation processes have often occurred in distributed, non-proprietary spaces, and been consumer-citizen driven, as epitomized by the open source software and publishing movements (Meikle, 2002). With industry speculation that an increasing number of the world's Internet users are estimated to rely upon the mobile phone for Internet access (Yahoo CEO Daniel Rosenweig, quoted in Elkin, 2005) it seems the user-pays culture of premium rate services looms as a challenge to the gift economy of the Internet. As the rapid development of SMS illustrates, this concern needs to be balanced against another major uncertainty in mobile communications, and that is whether consumers will follow carriers and manufacturers along these chains of added value charged at premium rates or whether consumer-citizen-led innovation will take the technology somewhere else altogether.

The premium rate services strategy proceeds from the expectation that, because mobile phone users have been habituated to make regular payments for connectivity, this will translate to a general willingness to pay for data services (Feldmann, 2005, p. 32). However, in many markets, these expectations remain unfulfilled. There are notable exceptions in particular countries and also with respect to specific applications. Games have proved to be highly successful in East Asian mobile markets in particular (*Screen Digest*, 2005; Rao, 2005). Many of these premium rate services blend peer-to-peer interaction with content. Such examples suggest that where the right balance is struck between contact, content and price, the premium rate services model can be very successful.

Ring tones, music downloads, and games are high-value applications for carriers and manufacturers. Although these can be lucrative and are deserving of further research as new circuits of popular culture, it is also the case that their success appears to be dependent upon heavy marketing (as the example of i-Mode indicates; cf. Ito *et al.*, 2005). In other words, the 'fit' of these services with the underlying demand-side trends in new media which favour more diverse modes of interaction is limited. The comparatively low-value data service of SMS voting has emerged as the premium rate services 'killer app', at least for the moment.

Cross-media SMS Applications: Retro-fitting Broadcast TV with Interactivity

Cross-media premium rate service applications have proved to be enormously popular (Feldmann, 2005), and particularly successful as a back channel or return path for broadcast media, especially television. In many places interactive television has been popularly realized not through digital television or pay television platforms, or other much-vaunted interactive television technologies, but through incorporation of the humble telephone in programme formats. The unproblematic invocation of interactivity as the distinction between 'old' and 'new' media has been soundly critiqued, and the multi-dimensionality of interactivity is now under extensive discussion (e.g. Manovich, 1995; Downes & McMillan, 2000; van Dijk & de Vos, 2001; McMillan, 2002; Kiousis, 2002). For the purpose of this analysis the cybernetic typology of interactivity, initially proposed by Bordewijk & van Kaam (1986) in the mid-1980s, provides a helpful starting point in understanding the significance of SMS interactivity for television programming. This framework of analysis was initially proposed as an aid to thinking about the implications of convergence for content regulation. It characterizes the predominantly one-way flow of broadcasting as a type of public address, or 'allocution'; the distributive flow of pay and subscription media as 'consultation'; the reciprocal flows of telephony as 'conversation'; and the information capture and data-mining capabilities of communications utilities (that is, telecommunications companies) as 'registration' systems. Importantly, all of these types of interaction can be mixed dynamically in digital systems, unlike their analogue antecedents.

Premium rate services were first integrated into a television programme format as the means for audience interaction by format production specialist Endemol, in *Big Brother*, in 2000 (Simmons, 2003). Consumer-citizens could call a telephone number to register their votes for characters they wanted to evict from the *Big Brother* house. To interact with the programme by voting, consumer-citizens paid a premium over and above the normal telephone call charge. In this way, the allocution of broadcasting address was adapted by structurally incorporating very rudimentary conversation and end-user registration into the programme format. Not only did this use of premium rate services offer the opportunity for consumer-citizen input to shape the narrative direction of the programme and thereby heighten end-user engagement, but it also generated a new revenue stream in which all interests in the premium rate service value chain could share, including broadcast licensees.

This cross-media application of SMS premium rate services has coincided with the rise of new 'reality TV' programming genres. It has also provided a means for renewing older light entertainment genres such as talent quests and quiz shows. These programmes codify new sets of cultural expectations, and viewing habits, in which telephony and data are crucially embedded. What rapidly emerged as television-plugged-into-telecommunications continues to evolve. Mobile text messaging is now a crucial platform, although the type of interactivity it supports in programme formats has not significantly extended beyond rudimentary registration and consultation. As we discuss below, further experimentation with conversational interaction has occurred using SMS and MMS in conjunction with programme formats. These developments still fall a long way short of the more complex types of interactivity that are characteristic of other emergent forms of cultural production, such as those that arise from the social networks associated with blogs or multi-user online computer games (e.g. Banks, 2002; Humphreys, 2005). Nevertheless, the adaptive strategy of embedding premium rate services into television formats is an interesting exploration of alternative development pathways for the mature mass medium, which is also transitioning to digital delivery platforms. Mobility has been crucial here because it has helped to greatly increase the popularity and accessibility of this type of engagement with broadcast media, especially in younger demographic groups which television intensely competes to lure away from other, expansive rich media and entertainment choices (Nightingale & Dwyer, 2006, p. 31). The earlier discussion of global adoption rates also suggests that mobile services significantly extend the global reach of interactive television applications far beyond that which is presently achieved by either the Internet or fixed-line telephony. These developments simultaneously address a deficit of interactivity of broadcast media, and allow for mass customization of television—that is, the customization of television content by a mass of end-users—as well as generating new revenue streams.

To take the Australian experience, in the first two series of *Big Brother* (2001/2002), the returns from premium rate voting were modest relative to total production costs but nevertheless important. For example, programme producers could apply their share to defray production costs, including prize money. The success of this model was quickly replicated in a range of other new programme formats including, for example, *Who Wants to be a Millionaire*. Although the audiences for the third series of *Big Brother*, broadcast in 2003, fell by 13 per cent from the previous series, eviction votes reportedly increased by 20 per cent, with 60 per cent of votes lodged by SMS. SMS revenues for this series were estimated at AUS$5.5 million. Half was retained by the carrier (Telstra) with the balance equally distributed between the production company Southern Star/Endemol, the broadcaster Network TEN, and the premium rate service intermediary Legion Interactive (Schulze & Sainsbury, 2003).

Many factors account for the success of the *Big Brother* format, including the global innovation system that is embedded within it (Cunningham, 2002, p. 11; Roscoe, 2004). This innovation system ensures that the format continues to evolve by feeding back intelligence on the success of new features tested through each iteration of the programme. This has included experiments with new modes of interactivity achieved via an increased variety of premium rate services applications. For example, when *Big Brother 3* went to air in the United Kingdom audiences were invited to send text messages to the programme, in addition to registering votes by SMS. There was an extraordinary response to this invitation to conversational engagement, with some 500,000 text messages received. These were moderated and retransmitted in a text 'strapline' on the bottom of the television screen. In this way, the living room conversation about *Big Brother* was expanded to a national scale, and then to a continental scale in the case of African *Big Brother*, which used the same feature with similar popular responses despite the low levels of telephone penetration in many parts of the continent (*All Africa News*, 2003; BBC Worldwide Monitoring, 2003).

In 2003, when *Big Brother 4* was broadcast in the United Kingdom, yet another form of premium rate interaction had been developed, relying on MMS. *Big Brother* fans could download picture alerts, ringtones, logos, video clips, java games, and animations, and they did, but only in modest numbers. The comparatively low response rate to this type of interactivity was interpreted by industry stakeholders to be a natural consequence of the modest number of MMS-capable handsets in use, and which have also been heavily promoted through *Big Brother* sponsorships. We suggest, additionally, that this type of consultational approach remains a very rudimentary use of the interactive potential of MMS, and is shaped more by the imperative of the commercial partnerships involved to blaze the shortest route possible to new revenue streams. There is little room in this service development context to pursue the directions that consumer-citizen-led innovation might take. MMS-capable handsets can support peer-to-peer distribution of promotional materials such as television clips released by broadcasters for circulation amongst viewers. More significantly, it has the potential to support the creation and circulation of user-generated content, but this is something yet to be fully explored by *Big Brother* and other premium rate services stakeholders. In one of its simplest forms, this includes pictures taken by camera-enabled mobiles that can be sent to other users.

Big Brother also illustrates the complexity of the premium rate service value chain. For example, in the case of services associated with a TV show, call revenues can be split between five or more parties: the carrier that provides carriage as well as billing and collection services, the premium rate service provider that provides number ranges and call management services, the broadcaster that promotes the service, the programme producer that provides content, and the programme format owner that has deployed premium rate services as a structural feature of the show. What cross-platform applications also indicate is that contact, not just content, is an important part of a successful premium rate service mix.

Above all else, this analysis of *Big Brother* shows that interactivity is a resource that is being carefully managed in the process of retrofitting television to remain relevant and commercially competitive in the digital networked era. Interactive techniques such as consultation and registration lend the impression of conversational interaction and direct participation, but in fact are substitutes for it. Instances of truly conversational interaction are more likely indicated in those disruptive moments when producers have indeed temporarily lost control to programme audiences or participants. It is this disruptive potential of interaction that stimulates the cultural studies interest in the programme and which further investigation could also reveal to be the kernel of popular interest in the programme.

Mobile interaction expands the range of ways in which interaction with television can be achieved. By no means the only way to interact with television, for a number of reasons mobile telephone platforms are nevertheless favoured in the premium rate services value chain over either fixed telephone networks or the Internet. These include the data capabilities of mobile phones, the higher levels of global accessibility and adoption, and the premium rate services business model. Just as important, however, is the fact that in these proprietary network spaces the parameters of interaction—for what type of interaction occurs and when—are generally not decided by consumer-citizens.

Concluding Comments: New Circuits of Culture

There is now an extensive scholarly literature as well as public discourse on mobile text messaging, and its incorporation into the cultures of everyday life. There is also a burgeoning debate as to the lessons to be learnt from SMS's popularity. There are cautionary notes being sounded, for instance, on the assumption that MMS can be regarded simply as the natural, and so lucrative, successor technology (Jenson, 2005; Scifo, 2005). What has been rather overlooked in mobiles scholarship, and discussions of what unfolds after the mobile phone, is the iterative, supplemental shaping of technology found in premium rate SMS services. Such adaptations and extensions of 2G mobile technology may certainly pale in comparison to the *grand récit* of 3G, yet the whole notion of a paradigm shift onwards and upwards in terms of data speeds, service capabilities, and communicative potential is questionable.

The figure of the consumer-citizen, their endless shaping of relatively 'closed' technologies such as mobiles through customization, and their co-creation of emerging mobile forms, genres, and content has been a clamorous theme in discussions of new media cultures (Green *et al.*, 2001; Hjorth, 2005a, b; Marshall, 2004; Oudshoorn & Pinch, 2003). As we have suggested, new media business models are increasingly organized around the recognition and incorporation of end-user involvement in the creation of services, including content. Scalable, consumer-citizen customization of media and communications, for example selecting a unique combination of mobile handset, ringtone and wallpaper and then using this device to participate in deciding the outcome of a talent quest, is only the beginning of this trajectory. Mobile data services are now a well-established, if under-studied, way for audiences to interact and be active with television stations and programme producers.

What we have highlighted here is the extent to which premium rate SMS services have become pivotal to international format developments in television, across a number of national and regional cultures and languages. Such developments merit detailed study, comparison, and further theorization. This work would be useful, not least at a time when mobile television in the form of DVB-H (digital video broadcasting—handheld) is being experimented with around the world, with accompanying discourses of cultural revolution (yet again) quickly modulating to descriptions of 'snack TV'; when the video iPod is in the marketplace; and when downloading and radical customization of television programmes and schedules is possible via peer-to-peer programmes such as Bittorrent. Much is claimed for such supposedly fundamental shifts in television and how it connects with its various publics, yet ironically the role of networked technologies such as the mobile phone, and now mobile media, in all this are under-studied and under-theorized.

At the very least, we would suggest that the widely used innovations in premium rate SMS and MMS, and their affiliated developments in modes of mobile media, demonstrate consumer-citizen interest in the capacity for interactive engagements with media, 'that is articulated through the audience re-making and re-forming cultural forms' (Marshall, 2004, p. 107). Prior to the digitization of communications, it might be said that this demand for interactive modes of communication was frustrated by the rigid control architectures of modern mass media. With the attachment of mobile devices and networks to television receivers, recording devices, and broadcasting networks there has been a circumscribed and narrow but definite and potentially significant opening for cultural expression and exchange—the future of which is uncertain.

Notes

[1] In fact only about 300,000 names were collected for the Live8 petition via SMS. A key reason for this was that the logistics of making a premium rate service available simultaneously in 130 countries were both significant and unprecedented (Grenville, 2005; Minick, 2005).

[2] We acknowledge the extremely helpful comments of our anonymous reviewers for this important clarification.

References

All Africa News (2003) 'Nigeria: more fun, more surprises and BBA enters third week', 11 Jun., LexisNexis.

Australian Broadcasting Commission (ABC) (2004) 'China's mobile phone wars', *The Buzz*, transcript, 7 Aug., Available at: http://www.abc.net.au/rn/science/stories/s1170716.htm (accessed 6 June 2005).

Banks, J. (2002) 'Gamers as co-creators: enlisting the virtual audience—a report from the net face', in *Mobilising the Audience*, eds M. Balnaves, T. O'Regan & J. Sternberg, University of Queensland Press, Sternberg, Brisbane, pp. 188–221.

BBC Worldwide Monitoring (2003) 'Africa: writer wraps "brainless hedonism" of reality TV show *Big Brother*', 13 Jul., LexisNexis.

Bordewijk, J. L. & van Kaam, B. (1986) 'Towards a new classification of tele-information services', *Intermedia*, M9, pp. 16–21.

Bruns, A. & Jacobs, J. (eds) (2006) *Uses of Blogs*, Peter Lang, New York.

CMI (2005) 'Mobile gaming: the market of the future?', *China Media Intelligence*, 2 Jun.

Cunningham, S. (2002) 'Culture, services, knowledge or "is content king, or are we just drama queens?"', in *Proceedings of Communications Research Forum*, Canberra, Department of Communications, Information Technology and the Arts, Available at: http://eprints.qut.edu.au/archive/00000202/ (accessed 5 Apr. 2007).

Cunningham, S. (2005) 'Creative enterprises', in *Creative Industries*, ed. J. Hartley, Blackwell, Malden, MA, pp. 282–298.

Downes, E. & McMillan, S. (2000) 'Defining interactivity: a qualitative identification of key dimensions', *New Media & Society*, vol. 2, pp. 157–179.

Economides, N. (1996) 'The economics of networks', *International Journal of Industrial Organization*, vol. 14, Available at: http://www.stern.nyu.edu/networks/site.html (accessed 5 Apr. 2007).

Elkin, T. (2005) 'Just one minute', *Media Post*, 30 Sept., Available at: http://publications.mediapost.com/index.cfm?fuseaction = Articles.showArticle&art_aid = 34694 (accessed 5 Apr. 2007).

Feldmann, V. (2005) *Leveraging Mobile Media: Cross-media Strategy and Innovation Policy for Mobile Media Communication*, Physica, Heidelberg and New York.

Goggin, G. (2006) *Cell Phone Culture: Mobile Technology in Everyday Life*, Routledge, London and New York.

Green, N., Harper, R., Murtagh, G. & Cooper, G. (2001) 'Configuring the mobile user: sociological and industry views', *Personal and Ubiquitous Computing*, vol. 5, pp. 146–156.

Grenville, M. (2005) 'Not so many Live8 SMS', 160 Characters SMS & Mobile Messaging Association, News, 11 Jul., Available at: http://www.160characters.org/news.php?action = view&nid = 1643 (accessed 5 Apr. 2007).

Häikö, M. (2002) *Nokia: the Inside Story*, trans. O.V. Virtanen Pearson Education, London.

Hartley, J. (1999) *Uses of Television*, Routledge, London and New York.

Hjorth, L. (2005a) 'Fast-forwarding present: the rise of personalization and customization in mobile technologies in Japan', *Southern Review*, vol. 38, pp. 23–42.

Hjorth, L. (2005b) 'Locating mobility: practices of co-presence and the persistence of the postal metaphor in SMS/MMS mobile phone customization in Melbourne', *Fibreculture Journal*, no. 6, Available at: http://journal.fibreculture.org/issue6/issue6_hjorth.html (accessed 5 Apr. 2007).

Humphreys, S. (2005) 'Productive players: online computer games' challenge to conventional media forms', *Journal of Communication and Critical/Cultural Studies*, vol. 2, no. 1, pp. 36–50.

Ito, M., Okabe, D. & Matsuda, M. (eds) (2005) *Personal, Portable, Pedestrian: Mobile Phones in Japanese Life*, MIT Press, Cambridge, MA.

Jakes, S. (2005) 'Li Yuchun: loved for being herself', *Time Asia*, 3 Oct., Available at: www.time.com/asia/2005/heroes/li_yuchun.html (accessed 5 Apr. 2007).

Jenkins, H. (2006) *Convergence Culture: Where Old and New Media Collide*, New York University Press, New York and London.

Jenson, S. (2005) 'Default thinking: why consumer products fail', in *The Inside Text: Social, Cultural and Design Perspectives on SMS*, eds R. Harper, L. Palen & A. Taylor, Springer, Dordrecht, pp. 227–298.

Keane, M. & Spurgeon, C. (2005) 'Advertising industries and China's creative vision', Paper presented to *Media and Society in Asia: Transformations and Transitions*, 14th AMIC Annual Conference, Beijing, 18–21 Jul.

Kiousis, S. (2002) 'Interactivity: a concept explication', *New Media & Society*, vol. 4, pp. 355–383.

Lindholm, C., Keinonen, T. & Kiljander, H. (eds) (2003) *Mobile Usability: How Nokia Changed the Face of the Mobile Phone*, McGraw-Hill, New York.

Manovich, L. (1995) *The Language of New Media*, MIT Press, Cambridge, MA.

Marshall, P. D. (2004) *New Media Cultures*, Arnold, London.

McMillan, S. (2002) 'A four-part model of cyber-interactivity. Some cyberplaces are more interactive than others', *New Media & Society*, vol. 4, pp. 271–291.

Meikle, G. (2002) *Future Active: Media Activism and the Internet*, Routledge, New York.

Middleton, C. (2002) 'Exploring consumer demand for networked services', Paper presented at the *Twenty-third International Conference on Information Systems*, Barcelona, 15–18 Dec., Available at: http://aisel.isworld.org/article_all.asp?Publication_ID = 31 (accessed 5 Apr. 2007).

Minick (2005) 'Minick powers global mobile petition for Live 8', Media Release, Available at: http://www.minick.net/press_live8.html (accessed 5 Apr. 2007).

Nightingale, V. & Dwyer, T. (2006) 'The audience politics of "enhanced" television formats', *International Journal of Media and Cultural Politics*, vol. 2, pp. 23–42.

O'Regan, T. & Goldsmith, B. (2002) 'Emerging global ecologies of production', in *The New Media Book*, ed. D. Harries, British Film Institute, London, pp. 92–105.

Odlyzko, A. (2004) 'Telecom dogmas and spectrum allocations', in *Wireless Unleased*, White Paper, Available at: http://www.wirelessunleashed.com/articles.html (accessed 5 Apr. 2007).

Oudshoorn, N. & Pinch, T. (eds) (2003) *How Users Matter: the Co-construction of Users and Technologies*, MIT Press, Cambridge, MA.

Rao, M. (2005) 'Wireless and mobile impacts on news and entertainment', in *Asia Unplugged: the Wireless and Mobile Boom in the Asia Pacific*, eds M. Rao & L. Mendoza, Sage, Thousand Oaks, CA, pp. 128–150.

Reed, D. P. (1999) 'Weapon of math destruction', *Context Magazine*, Spring, Available at: http://www.contextmag.com/setFrameRedirect.asp?src = /archives/200212/0ToC.asp (accessed 5 Apr. 2007).

Rheingold, H. (2002) *Smart Mobs: the Next Social Revolution*, Perseus, Cambridge, MA.

Roscoe, J. (2004) 'Watching *Big Brother* at work: a production study of Australian *Big Brother*', in *Big Brother International: Formats, Critics and Publics*, ed. J. J. Jones, Wallflower Press, London.

Schulze, J. & Sainsbury, M. (2003) 'The TV in your pocket', *The Australian*, Media Supplement, 17 Jul., p. B01.

Scifo, B. (2005) 'The domestication of the camera phone and MMS communications: the experience of young Italians', in *A Sense of Place: the Global and the Local in Mobile Communication*, ed. K. Nyíri, Passagen, Vienna, pp. 363–374.

Screen Digest (2005) 'Gaming on the move', *Screen Digest*, Jan., pp. 17–20.

Simmons, L. (2003) 'SMS TV textual relationships', *Revolution*, Mar., LexisNexis.

Spurgeon, C. (2005) 'Lovers and losers: mobile phone services advertising and the new media consumer/producer', *Journal of Interactive Advertising*, vol. 5, no. 2, Available at: http:/www.jiad.org/vol5/no2/spurgeon/ (accessed 5 Apr. 2007).

Toffler, A. (1970) *Future Shock*, Bodley Head, London.

Van Dijk, J. & Vos, L. de (2001) 'Searching for the Holy Grail: images of interactive television', *New Media & Society*, vol. 3, pp. 443–465.

Wang, J. (2005) 'Youth culture, music and cell phone branding in China', *Global Media and Communications*, vol. 1, pp. 185–201.

Wikipedia. (2006a) 'The Eurovision Song Contest', 30 Oct., Available at: http://en.wikipedia.org/wiki/Eurovision_Song_Contest (accessed 5 Apr. 2007).

Wilson, J. (2006) '3G to Web 2.0? Can mobile telephony become an architecture of participation?', *Convergence*, vol. 12, pp. 229–242.

INDEX